Robert Geroch

Topology

1978 Lecture Notes

With 112 Figures

MINKOWSKI
Institute Press

Robert Geroch
Enrico Fermi Institute
University of Chicago

© Robert Geroch 2013
All rights reserved. Published 2013

Cover: Lecture notes are often written in similar environments

ISBN: 978-1-927763-17-9 (softcover)
ISBN: 978-1-927763-18-6 (ebook)

Minkowski Institute Press
Montreal, Quebec, Canada
http://minkowskiinstitute.org/mip/

For information on all Minkowski Institute Press publications visit our website at http://minkowskiinstitute.org/mip/books/

Contents

1	Introduction	1
2	Sets in the Plane	3
3	Distance in the Plane	9
4	Properties of Boundary, Interior, Bounded, and Connected	25
5	Continuous Curves	35
6	Metric Spaces	43
7	Boundary, Interior, Bounded, and Connected in Metric Space	53
8	Theorems on Sets in Metric Space	67
9	Topological Space	71
10	Interior, etc. in Topological Spaces	81
11	An Example: Connected Sets with Boundary Attached	95
12	Compactness	97
13	Continuous Mappings of Topological Spaces	105
14	Applications of Continuous Mappings	115
15	Conclusion	123
	Appendix	129
	About the Author	153

1. Introduction

A key activity in mathematics is the attempt to capture certain intuitive ideas. By "capture", we mean to express these ideas in a form which is precise and clear, i.e., in a form which can be communicated without requiring that one's listener be instilled with all the nuances of one's own intuition. Two closely related activities are those of isolating and generalizing. One wishes to know exactly which features of some collection of mathematical objects are relevant to which conclusions, and how these various features are related in more general contexts.

One example of these activities involves the system of real numbers. One knows that one can, among other things, add numbers, and that this operation satisfies certain properties. [For example, $3+19 = 19+3; (3+19)+31 = 3 + (19 + 31)$.] These properties are, of course, intuitively clear, and indeed represent part of our mental picture of what "addition" means. The response of the mathematician to this state of affairs would be the following. What is really relevant here? The key elements are apparently a set (in this example, the set of real numbers), together with a specified operation on this set (in the example, addition of numbers). The "addition-ness" of this operation would be expressed by demanding certain properties (such as those above) of this operation. One thus "captures" addition of real numbers by introducing the notion of a set together with an operation, subject to certain properties. This "distilled essence of addition" is called an abelian group. One example of an abelian group is of course the numbers under addition. But it turns out that there are many others. For example, the non-zero numbers under multiplication (which operation, of course, has all the same properties as addition, e.g., $2 \times 3 = 3 \times 2; (2 \times 3) \times 7 = 2 \times (3 \times 7)$) also form an abelian group. There are also examples in geometry, for instance, the set of rotation about the origin in the plane, or the set of translations in Euclidean three-dimensional space. An abelian group, then, is just "the numbers under addition", but shorn of all extraneous elements (such as the issue of which numbers are larger than which others; which are written with straight and which with curved lines), and generalized to allow other examples. at least two benefits flow from, in this example, formulating the notion of an abelian group. First, one acquire a sense of a deeper understanding of what "addition" is all about.

The second is more practical. It turns out that many facts about addition of numbers are actually true more generally for any abelian group. Thus, one has only to establish such facts once – for the general abelian group – whence they will automatically be available within each particular example.

Mathematics proceeds on two levels: the intuitive, in which one forms a mental picture of the objects with which one deals and their relationships with each other, and the more formal, which consists of the definitions, theorems, and proofs. It is the interplay between these two levels – and in particular the translation from the first to the second – which is one of the essential and exciting features of mathematics.

It is our purpose here to see mathematics in action: how this interplay works, how mathematics operates and builds, how mathematicians think. This can apparently only be done by means of an example, and our example is to be the branch of mathematics called topology, Why topology? It has several attractive futures. First, this is an area in which all of us have already a rich and strong intuition, on which we shall be able to draw. Second, topology has the feature that interesting theorems are very near the surface. That is to say, topology has a particularly high ratio: (what comes out as theorems) / (what has to be put in as definitions, etc.). A third attractive feature is that topology is very different from the mathematics with which one may be familiar, e.g., algebra, and so one is provided the opportunity to see the breadth that is mathematics. Finally, topology is attractive because it is an active field of current research in mathematics.

2. Sets in the Plane

In this section we do two things. First, we shall try to indicate, by means of some examples, the kind of structure one is (and the kind one is not) trying to capture in topology. That is, we indicate what topology is all about. Second, we shall discuss the intuitive meaning of certain notions which are, ultimately, to be defined, i.e., to be made precise in topology. Both of these goals are to be accomplished by a discussion and description of certain sets in the plane.

Recall the Euclidean plane, i.e., the space of plane geometry. We may represent points in the plane by the usual Euclidean coordinates, x and y. That is, we introduce orthogonal x- and y-axis, as shown on the right. Than any point in the plane is represented, by two numbers, its x- and y-coordinates as follows. First, drop perpendicular from the point to the x- and y-axes, as shown. Then the x-coordinate of the point is the distance from the origin to the foot of the perpendicular along the x-axis, and similarly for the y-coordinate. The point in the figure, then, would be represented (x, y). Any point in the plane is represented in this way by two numbers, and, conversely, any two numbers represent a unique point of the plane.

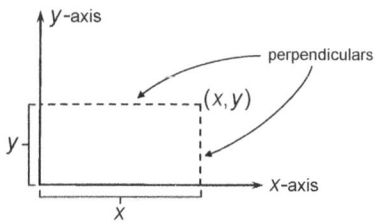

We shall be interested, for the moment, in various sets in the plane. To characterize a set, one must give some clear, unambiguous prescription for deciding which points of the plane are in the set and which are not in the set. This can be done in a number of different ways. For example, one could give an equation in the coordinates (x, y), e.g., "the set of all points (x, y) in the plane for which $y = x^2$". This is, of course, such a prescription: To decide whether or not a given point, rep-

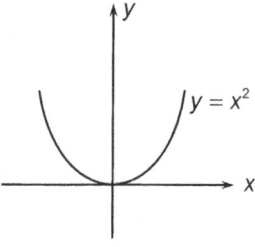

3

resented (x, y), is in this set, one computes x^2, and checks to see whether or not it is equal to y. The resulting set is of course that represented by the parabola in the figure.

One could also specify a set by giving an inequality in the coordinates, e.g., "the set of all points (x, y) for which $x^2 + y^2 < 1$". This set is of course the disk shown. More generally, any combination of equations and inequalities on (x, y) will provide a clear, unambiguous prescription, and thus will define a set in the plane.

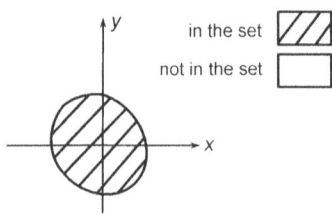

For example, "the set of all points (x, y) for which $x^{11} - 13/6\, x^3 y \le .16 y^5$ or $y^3 - xy \ne 0$, except that all points with $y = 3$ are excluded" describes a set in the plane, for again, given a candidate, (x, y), one could decide whether or not it is in the set. Sets can also be specified by other than algebraic equations. For instance, "the set of all points (x, y) for which x is a rational number" describes a set. This set, for example, consists of a collection of vertical lines in the plane, one for each rational x-value. Thus, there are vertical lines at $x = 1/2$ and at $x = -237/359$, but none at $x = \pi$, for π is of course not rational. As other examples, "the set of all points (x, y) for which the decimal expansion of y contains no digit 7" or "the set of all points (x, y) for which the area of that part of the circle of radius x^2 and center (x, y) which lies to the left of the y–axis is less than 3" specify sets in the plane.

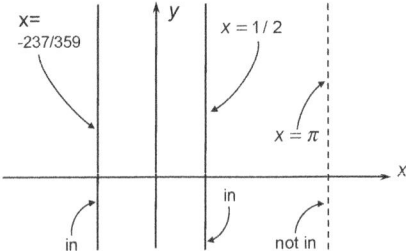

The following are examples of things which would not be regarded as specifying sets in the plane: "If you want to know whether (x, y) is in the set, flip a coin: heads, in; tails, out." "Point (x, y) is in the set if y is a positive integer, and the Bears win Super Bowl y." These would be regarded as unclear and/or ambiguous.

I hope these few examples make the distinction clear. In fact, there is an active branch of mathematics which deals in part with the issue of what, precisely, "specify a set" should mean.

We next wish to discuss certain intuitive features of various sets in the plane. We emphasize that the purpose of this discussion is merely to develop some common ground for later use. What follows is *not* mathematics.

Given a set in the plane, we wish to mean by its boundary the set of all points "on the edge of the given set". Consider, for example, the disk, the set of points (x, y) with $x^2 + y^2 < 1$. Its boundary should be circle shown, i.e., the set of points (x, y) with $x^2 + y^2 = 1$. This circle is also the boundary of the "disk with its edge included", i.e., of the set of points (x, y) with $(x^2 + y^2 \leq 1$. For the more complicated set illustrated in the second figure, its boundary would again be its edge, as indicated. Given a set in the plane, we wish to mean by its interior the set of all points which are "actually inside the set, i.e., neither external to the set nor on

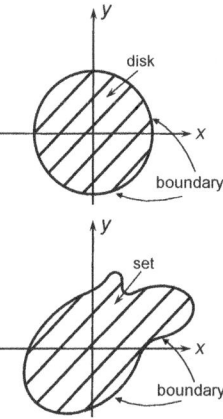

its boundary". Thus, for the two examples illustrated above, the interiors are to be the cross-hatched regions, excluding in each case the boundary.

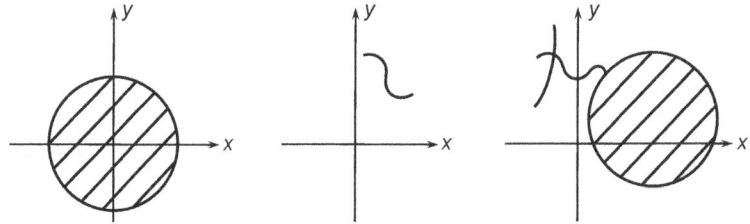

Thus, we have two intuitive features of sets, the boundary and interior. Each is to assign to a given set another set which is its interior. Each is to assign to a given set another set (i.e., to a given set the set which is its boundary), or the set which is its interior). The next two features will have a somewhat different character: They will be properties which a given set is to either have or not have. We wish to describe a set as being connected if it "consists of just one piece, not of two or more". As examples, the sets illustrated above (a disc, a segment of a curve, a disk with a segment of a curve going off of it and another segment of a curve crossing it) are to be regarded as connected. By contrast, the sets below (two disks, a disc and a point, a curve with a point missing) are not to be regarded as connected, for each "consists of two or more pieces".

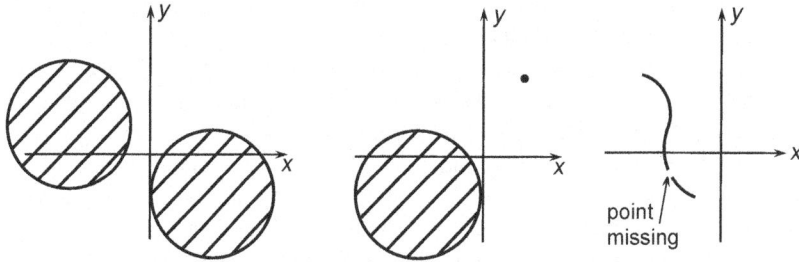

Finally, we wish to regard a set as bounded (nothing to do with "boundary") if it "remains in a finite region of the plane; does not go off to infinity". As examples, all of the sets illustrated on the previous page were bounded. Some examples of sets which are not to be regarded as bounded are the entire plan ("the set of all points (x, y)"), and an infinitive straight line (for example, "the set of all (x, y) with $x = 1$").

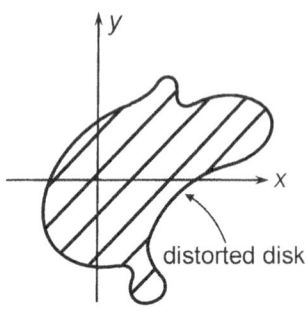
distorted disk

The four features described above, boundary, interior, connected, and bounded, will, as it turns out, all be incorporated into topology. By way of indicating in a general sense what topology is about, we can also give some examples of "features of sets" which will have no place in topology. Such features include "being a disk", "being a straight line", "being more than one foot in size". What is it about these features which will make them "non-topological"? It is a distinction which is difficult to make clearly at this point. All of these features refer to how the points of the plane are arranged to form the plane. The essence of the distinction is the following. Imagine that our sets are drawn, not on the rigid plane we have been imagining, but rather on a rubber sheet. We now allow our rubber sheet to be stretched and pulled (but not folded or torn), and ask which of our features thereby remain the same. For example, "being connected" does remain the same. A disk, for instance, is connected. If we were to draw our disk on a rubber sheet, and then stretch and pull, there might result (when the stretched and pulled sheet is imprinted back on the rigid plane) the set illustrated on the right. But this set (a "distorted disk") is of course also connected.

More generally, one convinces oneself that such a distortion (by imprinting on a rubber sheet, stretching and pulling, and then imprinting back onto a rigid plane) always leaves any connected set connected, and any set not connected not connected. In a similar way, the boundary remains the same under our distortion. What "remains the same" means, in this case, is the following. Fix a set (say, a disk), and determine its boundary (in this example, its circular edge). Now imprint both the set and its boundary (say, with the set in blue and the boundary in orange, to keep them separate) on a rubber sheet. Stretch and pull the sheet, and then imprint both colors back onto the rigid plane. Then, as seems clear, the boundary of the resulting blue set will be just the resulting orange set. Similarly, interior and bounded remain the same under this stretching and pulling. But note, by contrast, that our non-topological properties are altered by this operation. That is to say, a set which is a disk, which is a straight line, or which is less than one foot in size, need not remain so under stretching and pulling of the rubber sheet. We thus have at least a rough criterion for what is likely to be a topological feature of sets in the plane and what is not. Topology is sometimes called "rubber sheet geometry".

Two other possible features of sets which will play no role in topology are "being excessively large" and "being attractive". For these, one does not perhaps have a sufficiently clear intuitive notion – or at least none which is likely to be in reasonable agreement with others – to even get started. It would be very difficult to pin down what it is one is trying to capture.

So, we have decided to focus on four features of set in the plane: boundary, interior, connected, and bounded. Why did we call our treatment of these features an "intuitive discussion"? Why did we not instead regard "The boundary of a set is its edge.", "The interior of a set consists of the points inside the set.", "A set is connected if it consists of no more than one piece.", and "A set is bounded if it remains in a finite region of the plane." as the *definitions* of these four terms? The reason is that these sentences are not sufficiently precise to be regarded as proper definitions in mathematics. A definition in mathematics is supposed to be unambiguous, i.e., such that someone, on reading the definition but without using subtle nuances of what words mean in English, can decide in every case whether or not the conditions of the definition are applicable. But our characterizations of "boundary", "interior", "connected", and "bounded" are ambiguous. Some examples will illustrate this point. Consider the set in the plane consisting of the single point (0, 0), i.e., the point at the origin. What is the boundary of this set? One may have an opinion on this question (presumably, either that the boundary consists of no points, or that the boundary is just the single point at the origin), but I would venture that one is not so confident that others will share that opinion as one was, say, with the disk. What is the interior of this set? Is it this one point or no points. Consider, as a more extreme example, the set illustrated in the second figure on page 4. What is the boundary of this set? Is it all points in the plane? Just those (x, y) with x rational? Those with x irrational? None of the points in the plane? What is the interior of this set? Is this set connected? Clearly, our discussion of the terms "boundary", "interior", "connected", and "bounded" did not provide a clear prescription for deciding what the boundary or interior of a set is, whether a set is connected or bounded. Our discussion, in short, was not a definition in the sense of mathematics.

We see that a definition is a somewhat different thing in mathematics than in everyday English. [It is too bad that a single word is used for both.] For example, my dictionary defines "chair" as "a seat, usually movable, for one person". Clearly, this definition is intended merely to give one a general idea of what sort of objects are conventionally called chairs and what sort are not. It is certainly not intended to permit one to make, unambiguously for every object one sees throughout one's life, the decision as to whether or not that object is a "chair". In particular, "usually ..." could never be inserted in a definition in mathematics. We shall see some examples of mathematical definitions shortly.

3. Distance in the Plane

We now have before us a typical example of the "raw material" of mathematics. We have some intuitive notions (namely, that of the boundary and interior of a set; that of whether a set is connected or bounded), and also a general idea of the limits of our intuition. The task we face, as mathematicians, is to "capture" these notions, i,e,. to formulate some precise definitions, within mathematics, of the words "boundary", "interior", connected", and "bounded". We remark that we are not exactly trying to "discover" the definitions (as though they already exist, somewhere out there, and we are merely trying to bring them to paper) – but rather we must "concoct" them. The only requirements on our concocted definitions are, first, that they be clear and unambiguous, and, second, that they seem to reflect our intuitive ideas of what these words mean. In this section, we make a preliminary attempt at such definitions. This attempt will be refined shortly.

The first decision we face – a critical one – is that of what structure we shall use in the formulation of our definitions. The plane, of course, has a great deal of structure: that of "straightness" of lines, of "orthogonality" of intersecting straight lines, of "circle-ness" of circles, etc. But presumably only a small part of all this structure is 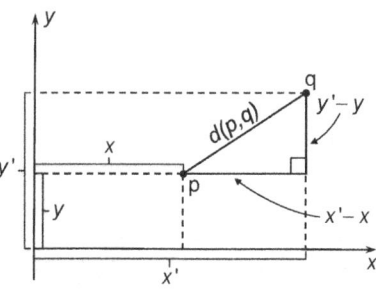 relevant to the sorts of things we are after. What we want to do, then, is somehow isolate what it is about the plane which gives rise to our ideas about "boundary", etc. In practice, one would normally try several different structures, selecting them initially based on one's judgment as to what looks promising, and finally based on what choices actually lead to appropriate definitions. [How accurate one's judgment is on such issues is one of the important distinctions between a good and a poor mathematician.] We, however, do not have the time to explore all of these false starts (e.g., to try "straightness", decide that it seems very difficult to formulate a definition for "boundary" in terms of straightness, and then finally realize that straightness was not a very good choice in the first place, since of course stretching and pulling of the rubber sheet certainly destroys straightness). We, therefore, merely jump to that structure which, as it turns out, works. Recall the notion of the geometrical distance (e.g., as measured by a ruler) between two points in the plane. We denote the distance between points p and q by $d(p, q)$. There is, of course, a simple formula for this distance in terms of the coordinates for the points. Let p have coordinates (x, y), and q coordinates (x', y'). Then, as shown in the figure, one constructs a right triangle, one leg of which has length $(x' - x)$, the other, length $(y' - y)$. But the hypotenuse of this triangle is the line segment joining p and q – a segment whose length we have agreed to denote $d(p, q)$. So, by the Pythagorean theorem, we have $d(p, q)^2 = (x' - x)^2 + (y' - y)^2$. Taking the square root of both sides, we obtain our formula for $d(p, q)$:

$$d(p, q) = \sqrt{(x' - x)^2 + (y' - y)^2}$$

Thus, for example, the distance between the point labeled $(1, 2)$ and the point labeled $(4, 6)$ is $\sqrt{(4 - 1)^2 + (6 - 2)^2} = \sqrt{3^2 + 4^2} = \sqrt{9 + 16} = \sqrt{25} = 5$.

Our program, then, is to give expression to our ideas about "boundary", etc., using this distance.

We begin with the notion "boundary". Let us, to fix ideas, focus on one particular set A, say the unit disk centered at the origin, i.e., the set of all points (x, y) with $x^2 + y^2 < 1$. Consider some point p, such as the point $(2, 2)$, as shown in the figure. Now we clearly do not want this point p to be in the boundary of our set A.

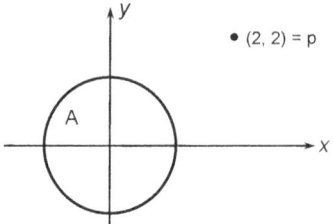

The question is: What is the basis for this attitude toward the point p? One might answer by saying that the distance of the point p from the set A is greater than zero. But this, unfortunately, is not a very useful expression of our attitude, for we do not know what "distance of a point from a set" means (only "distance of a point from a point"). Alternatively, one might say that the distance of the point p from the point at the origin (a distance, in this example, of about 2.8) is greater than 1, the radius of the disk A. Although this statement certainly does say what we want to say for this particular example, it is, unfortunately, rather strongly tied to the example itself (i.e., to the fact that A is a disk). If, for example, A were some more complicated set, then we would not be able to say "... the radius of the disk A. What we want to say, of course, is that our point p should not be on the boundary of the disk because it is "well away" from the disk. After a few more similar tries, one might hit upon the following statement: There is no point within distance 1/2 of p which is also in A. This statement is, of course true. Indeed, the locus of points within distance 1/2 of p is just the disk of radius 1/2 centered at p, and this disk does not intersect the set A, i.e., it has no points in common with A. So, one might be tempted to regard a point p as "well outside" of set A (so, not on the boundary of A) if there is no point within distance 1/2 of p which is also in A. But this, unfortunately, is still not quite right.

Consider, for example, the point p' of the figure. Now this p' is certainly "well outside" the disk (perhaps not as "well outside" as was p, but at least "well enough outside" so that we do not want p' to be on the boundary of A). However, there *is* a point within distance 1/2 of p' which is also in A, namely the point indicated in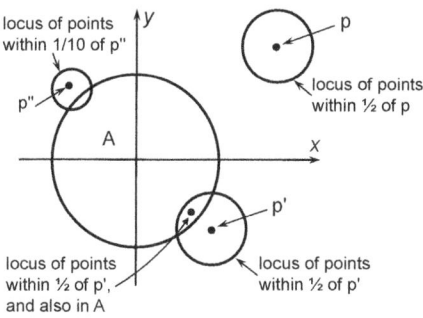
the figure. So, one might think of changing our statement to: Point p is well outside of set A if there is no point within distance 1/10 of p which is also in A. This statement of course holds for both our point p and our point p'. But this will not work either. Consider the point p'' in the figure. It is still to be "well outside" of A, but now there *is* a point within distance 1/10 of p'' and also in A. Our mistake is that we are committing ourselves to "how well outside" (i.e., to the distance, 1/2 or 1/10) *before* we know which point (p or p' or p'') we are to consider. All we have to do is reverse the order of commitment (i.e., allow ourselves to choose the distance *after* we know the point to be considered). This we do by the following statement: We regard a point p as "well outside" of A if, for some positive number ϵ, no point is within ϵ of p and also in A. [Note the word "some": We get to choose ϵ, the distance, after we know what p is.] This statement works. Consider, for example, the figure above. The point p, according to this statement, is well outside of A, for we may choose $\epsilon = 1/2$: There is no point within 1/2 of p which is also in A. That is, there is *some* positive ϵ (namely, 1/2) such that no point within ϵ of p is also in A. The point p' is also "well outside" of A, for *some* positive ϵ (namely, 1/10), there is no point within ϵ of p' which is also in A. Finally, the point p'' is also "well outside" of A, e.g., by choosing $\epsilon = 1/100$.

All we have done so far is identify certain points which we do not want to be in the boundary of our set, namely those which are "well away" from the set, in the sense above. But we also want to exclude from the boundary those which are "well inside" the set. Consider, for example, the point p in the figure, given, say by (1/2, 1/2). This p, of course, is not to be on our boundary. How do we express the fact that it is "well inside" our set A? Again, one would try various criteria, most of which would not look too promising. [E.g., the distance of p from the origin (about .7 in this example) is less than the radius of the disk. But this criterion depends on our particular set A.]

Finally, one would notice that the "well inside-ness" of p can be expressed thus: Every point within distance 1/4 of p is also in A. Indeed, the locus of points within distance 1/4 of p is the small disk shown, and all such points are in our set A. Again, however, the criterion "all points within distance 1/4 of p be also in A" turns out not to be quite right. The point p' of 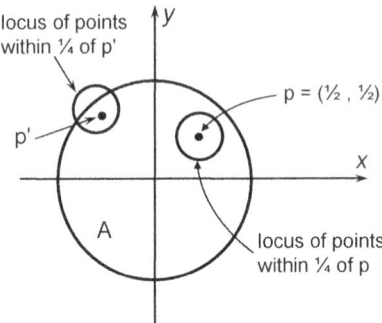 the figure, which we would certainly wish to regard as being "well inside" A, does not have the property that all points within distance 1/4 of p' are also in A. Again, we must allow ourselves to choose "how well within", i.e., to choose the distance, *after* we know which point we are to consider. Thus, as before, end up with the following: We regard point p as "well inside" set A if, for some positive ϵ, every point within ϵ of p is also in A. With this criterion, then, the point p in the figure above is "well within" A (choosing $\epsilon = 1/4$), as is the point p' (now choosing, say $\epsilon = 1/100$).

We now have all the ingredients to formulate our definition. We want the boundary of A to be what is left over after one excludes all the points which are "well outside" A, in the sense above, and also all points which are "well inside" A, in the sense above. All we have to do is figure out how to say all this in one sentence. One convinces oneself (e.g., by reexamination of the figure on page 11) that point p fails to be "well outside" of A provided that, for every positive ϵ, there is a point within ϵ of p and also in A; and that point p fails to be "well inside" A provided that, for every positive ϵ, there is a point within ϵ of p and also outside of A. But the boundary of A is supposed to be just those points which both fail to be "well outside" of A and fail to be "well inside" A. Thus, we arrive at:

<u>Definition</u>. Let A be a set in the plane. Then the *boundary* of A is the set of all points p such that, for every positive number ϵ, there is a point q with $d(p, q) \leq epsilon$ and q in A, and also a point q' with $d(p, q') \leq \epsilon$ and q' outside of A.

A few remarks are in order. First, the words which follow "Definition" above are the first things we have written which may be regarded as "real mathematics". All else – all the discussion, false attempts, examples, and so on – was merely some general remarks to motivate and get one in the mood. What preceded the definition, on the other hand, is a good example of the sort of activity in which mathematicians are engaged. [Only a small fraction of the time of a mathematician is normally spent "calculating", or "working with formulae".] Good judgment is no less important an attribute for a mathematician than for anyone else. The second remark involves again the issue of what a definition is supposed to do in mathematics. The role of this definition is to permit one to decide, unambiguously, which points are to be in the boundary of any given set A and which are not. This is quite a different role from that of the definition of "chair". To decide whether or not an object is a "chair", one may use the dictionary definition as a rough guide, but one also uses one's lifetime of experience with chairs. The dictionary definition is not intended to replace that experience. But not so in mathematics. The following point, which sounds trivial when stated, but which seems often to be overlooked in the heat of the moment, cannot be overemphasized: *To decide what the boundary of a set is, one uses precisely what it says in the definition, nothing more and nothing less.* One does not use one's ideas or thoughts about boundaries; one does not think. Indeed, this is why we had to work so hard to get the definition. This is one feature which sets mathematics apart from other disciplines.

We shall denote the boundary of set A by bnd (A).

We now give some examples of application of the definition to determine the boundaries of various sets. Let us consider first the set we have just used as our example, the disk. Is the point p in bnd (A)? Is it true that, for every positive ϵ, there is a point q within ϵ of p and in A, and also a point q' within ϵ of p and outside of A? The answer is again no. Set $\epsilon = 1/2$. Then there is no point q within distance ϵ of p and in

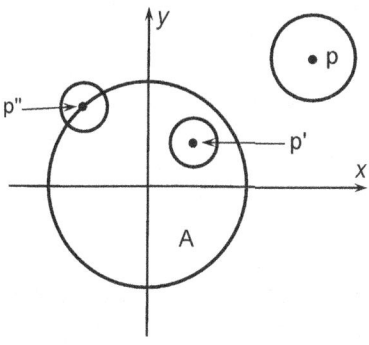

A. Indeed, the locus of points within $1/2$ of p is the disk shown. Is p' in bnd (A)? Is it true that for every positive ϵ, there is a point q within ϵ of p and in A, and also a point q' within ϵ of p and outside of A? The answer is again no. Set $\epsilon = 1/4$. Then there is no point q within distance ϵ of p and outside of A, for the disk shown lies entirely within A. Finally, we consider the point p''. Is p'' in bnd (A)? Is it true that, for every positive ϵ, there is a point q within ϵ of p and in A, and also a point q' within ϵ of p and outside of A?

The answer, we claim, is yes. Consider, for example, $\epsilon = 1/2$. The locus of points within distance $1/2$ of p'' is the disk shown. Is there within this disk a point in A, and also in this disk a point outside of A? There certainly is, namely the two points shown.

Consider, then, another value for ϵ, say $\epsilon = 1/10$. Then the locus of points within distance $1/10$ of p'' would be a smaller disk. Is there within this disk a point in A, and also in this disk a point outside of A? Again, there is. Clearly, then, for *every* positive ϵ, there is a point within ϵ of p'' and in A, and also a point within ϵ of p'' and outside of A. That is to say (recalling the definition), the point p'' is on the boundary of A. It should now be clear that bnd (A), the boundary of A, is just the circular "edge" of A, that is, the set of all points (x, y) with $x^2 + y^2 = 1$.

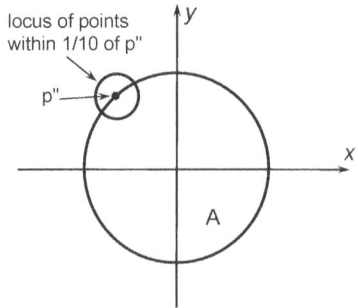

Suppose next that we let A instead be the disk with its boundary, i.e., the set of all points (x, y) with $x^2 + y^2 \leq 1$. Repeating the previous paragraph word for word, we find that bnd (A) is the same as the set above, namely the set of all points (x, y) of the plane with $x^2 + y^2 = 1$.

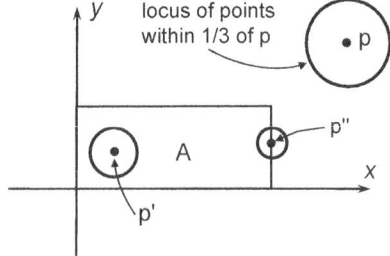

Next, let A be the rectangle in the plane given by all (x, y) with $0 < x < 2$ and $0 < y < 1$. We determine bnd (A). Just as above, the point p is not in bnd (A) (for it is not true that, for every positive ϵ, there is a point within ϵ of p and in A, and also a point within ϵ of p and not in A. Specifically, choose $\epsilon = 1/3$ as shown. Then no point within $1/3$ of p is in A.) Similarly, the point p' is not in bnd (A). The point p'', however, is in bnd (A), for, given any positive ϵ, there is indeed a point within ϵ of p'' and in A, and also a point within ϵ of p'' and outside of A. Thus, the boundary of this set A is again its "edge". More precisely, bnd (A) is the set of all points (x, y) satisfying either i) $x = 0$ and $0 \leq y \leq 1$, or ii) $x = 2$ and $0 \leq y \leq 1$, or iii) $y = 0$ and $0 \leq x \leq 2$, or iv) $y = 1$ and $0 \leq x \leq 2$. [These four choices define the four "sides" of the rectangle.]

We now modify this last example slightly. Let A now be "two squares side by side, but without the line between them included", i.e., the set of all (x, y) with $0 < y < 1$ and either $0 < x < 1$ or $1 < x < 2$. Is the point p shown (e.g., $(1, 1/2)$) in bnd (A)? Is it true that, given any

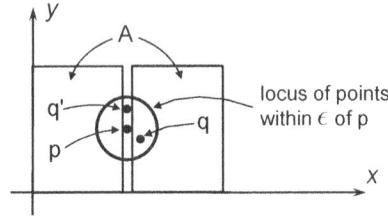

positive ϵ, there is a point within ϵ of p and in A, and also a point within ϵ of p and outside of A? The answer is yes. The locus of all points within ϵ of p, for a typical ϵ, is shown. There is indeed a point (namely, q in the figure) within ϵ of p and in A, and also a point (namely, q' in the figure) within ϵ of p and outside of A. Clearly, this holds for any ϵ. So, our point p is in bnd (A).

We consider next a somewhat more exotic example. Let A be the line given by: the set of all (x, y) with $y = 1$. We determine bnd (A). The point p, for example, is not in bnd (A), for it is not true that, for any positive ϵ, there is a point within ϵ of p and in A, and also a point within ϵ of p and outside of A. Choose, say, $\epsilon = 1/4$, as shown.

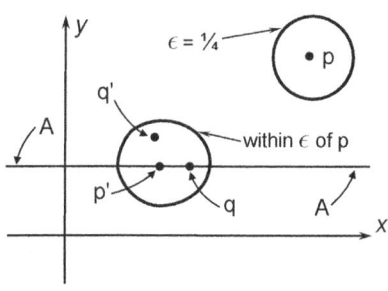

Then *no* point within $1/4$ of p is in A. The point p', by contrast, is in bnd (A). Given any ϵ, choose q (within ϵ of p and in A), and q' (within ϵ of p and outside of A) as shown. We conclude, then, that the boundary of this set A is just the set A itself.

Let A be the single point at the origin, $(0, 0)$. Is this point itself in bnd (A)? It is. Is it true that, given any positive ϵ, there is a point within ϵ of p (the origin) and in A, and also a point within ϵ of p and outside of A? There certainly is: For the point within ϵ of p and in A, choose p itself (clearly within ϵ of p, and certainly in A); for the point

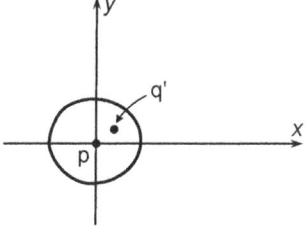

within ϵ of p and outside A, choose the point indicated q' in the figure. Thus, for this set A, we have bnd $(A) = A$.

Let A be entire plane, i.e., the set of all points (x, y) in the plane. What is bnd (A)? Let's pick a point p, say the origin, and try it. Is it true that, given any positive ϵ, there is a point within ϵ of p and in A, and also a point within ϵ of p and outside of A? There is not, and in fact there is no point in the plane "outside of A", for now A is the entire plane. So, the origin p is not in bnd (A). Similarly, any other point p we choose will not be in bnd (A). Thus, in this example bnd (A) is the set having no points. [It is customary to express our conclusion in this way, and *not* "There is no bnd (A).] The set having no points, i.e., the set such that for all (x, y) this point is not in the set, is called the *empty set*. [Note that this is an unambiguous statement of what is in and what is out of the set, namely, everything is out.] Our conclusion, then, is that the boundary of the entire plane is the empty set.

Let A be the empty set. What is bnd (A)? Pick a point p, say the origin

(0, 0). Is it true that, given any positive ϵ, there is a point within ϵ of p and in A, and also a point within ϵ of p and outside of A? It is not true, and in fact there is no point whatever in A", since now A is the empty set. So, this p is not in bnd (A) – and clearly no other point will be in bnd (A) either. We conclude, then, that the boundary of the empty set is the empty set.

These examples are intended, not only to illustrate what our definition of the boundary gives under various circumstances, but also to make the following point: When confronted with a definition in mathematics, one should, in the first instance, apply it mechanically to various examples, and see what it gives. Eventually, after one has "experimented" with enough examples, one acquires a feeling for how the definition works, and in particular one becomes able to apply the definition without going through all the words every time. If you do not feel that at this point you can do this for "boundary", I would suggest that you make up some sets of your own, and apply the definition to determine their boundaries.

We next turn to the definition of the second of our notions, that of the interior. Recall that we would like the interior of a set to consist of the points "well inside" that. But we have already obtained, on page 11, a precise formulation of this intuitive notion. We are thus led immediately to:

Definition. Let A be a set in the plane. Then the *interior* of A is the set of all points p such that, for some positive number ϵ, every point q with $d(p, q) \leq \epsilon$ is in A.

Again, we apply the definition to various examples. We shall denote the interior of set A by int (A).

Let A be the disk, as shown in the figure on page 13. Is the point p in int (A)? Is it true that, for some positive ϵ, every point within ϵ of p is in A? The answer is no: Indeed, for any positive ϵ, the point p itself is certainly within ϵ of p, but is not in the set A. Is p' in int (A)? Is it true that, for some positive ϵ, every point within ϵ of p' is in A? Now, the answer is yes. Indeed, choose $\epsilon = 1/4$ (certainly *some* positive ϵ). Then the locus of all points within $1/4$ of p' is the disk shown in the figure, and clearly every point within this disk (i.e., every point within ϵ of p') is in A. So, this p' is in int (A). Is p'' in int (A)? Is it true that, for some positive ϵ, every point within ϵ of p'' is in A? The answer is now no. The locus of points within ϵ of p'', for a typical ϵ, is shown in the figure. But there is, within this small disk about p'', a point not in A, namely the point indicated in the figure. Clearly, this will hold for every ϵ. So, it is not true that, for some ϵ, every point within ϵ of p'' is in A. So, p'' is not in int (A). We conclude, then, that int (A) is just the disk itself, i.e., the set of all points (x, y) within $x^2 + y^2 < 1$. Note that this is of course the answer we wanted, i.e., that our definition seems to do what it is supposed to.

The interior of the rectangle A shown in the second figure on page 14 is just A itself.

Consider next the "two squares" shown in the last figure on page 14. Is p in int (A)? Is it true that, for some positive ϵ, all points within ϵ of p are in A? The locus of points within ϵ of p, for a typical ϵ, is shown in the figure. But it is not true that every point in this disk is in A, for example, the point q' shown in the figure. So, p is not in int (A).

Next, let A be the line as in the first figure on page 15. Clearly, p is not in int (A). What about p'? It is not either. Indeed, for any ϵ (a typical one shown in the figure), there is a point, such as the q' shown, which is written ϵ of p but which is not in A. So, p' is not in int (A). Clearly, then, there are no points in int (A). That is to say, int (A) is the empty set.

For A the single point, as in the second figure on page 15, int (A) is also the empty set.

What is the interior of the entire plane? Consider the point p at the origin. Is it true that, for some positive ϵ, every point within ϵ of p is in A? It sure is: Choose any ϵ you want (say, $\epsilon = 187$), because *every* point in the entire plane is in A. Clearly, the interior of the entire plane is the entire plane.

The interior of the empty set is the empty set.

Finally, let us consider the "disk with a hole in it", i.e., the set A given by all points (x, y) with $x^2 + y^2 < 4$ and $x^2 + y^2 > 1$. [The second condition "removes the hole".] Is the point p at the origin in int (A)? It is not, for it is not the case that, for every positive ϵ, all points within ϵ of p are in A. Indeed, for $\epsilon = 1/2$, there is not even *any* point within ϵ of p and in A. This example suggests, then, that one should think of

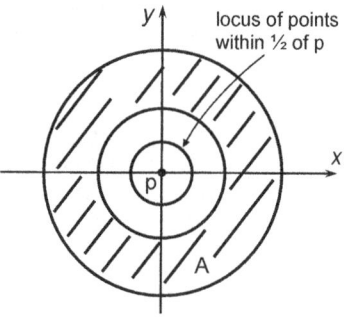

the interior as consisting, not so much of the points which are "surrounded" by A, but rather of the points which are "overrun" by A..

We turn next to the intuitive notion "bounded". Recall that bounded is supposed to mean "does not go off to infinity; remains in a finite region". How are we to express this precisely? What we want to say, roughly speaking, is that "the points of A do not keep getting farther and farther away from the 'bulk' of A". After a few false starts, one realizes that this is essentially the same as demanding that "the points of A do not keep getting farther and farther away from some point p of the plane."

But this idea, finally, can be expressed as follows.

<u>Definition</u>. A set A in the plane is said to be *bounded* if, for some point p of the plane and some positive number c, all points q of A satisfy $d(p, q) \leq c$.

What the definition requires, in other words, is that A lie entirely within some disk (with center p and radius c).

We give some examples.

Let A be the usual disk, i.e., the set of all points (x, y) with $x^2 + y^2 < 1$. This set A is bounded. Indeed, choose, say, p the point $(1, 1)$, as shown, and let $c = 4$. The locus of points within distance 4 of p is indicated in the figure. Clearly, all points of A are in this region, i.e., all points of A are within distance ϵ of p. So, A is bounded. [Clearly,

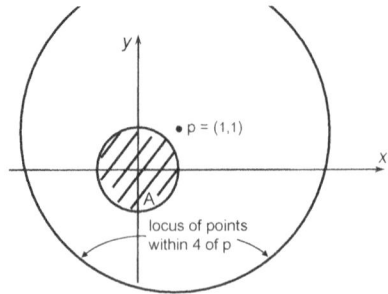

many choices of p and c are possible to show that A is bounded. We could, for example, have chosen $p = (0, 0)$, and $c = 1$, or $p = (20, 20)$, and $c = 50$.]

Similarly, the rectangle (the set represented by the second figure on page 14), and the "two squares" (the set represented by the last figure on page 14) are bounded. The set consisting of a single point (say, the origin) is bounded. Choose, say, $p = (0, 0)$, and $c = 738$. Then clearly all points of A (namely the one point which is in A, the origin) is within 738 of p.

The empty set is bounded. Choose anything you want for p and c (say, p the point $(-13, 4)$ and $c = 2/3$). Then every point in A (the empty set) is within $2/3$ of p, because there are no points in A. The entire plane is not bounded. Is it true that, for some point p and some positive c, all points of A (the entire plane) are within c of p? Hardly, for A is the entire plane, so if all points of A are to be within c of p, then all points of the plane would have to be within c of p. But for no choice of p and c are all points of the plane within c of p. That is, one cannot find p and c such that all points of A are within c of p. That is, our set A, the entire plane, is not bounded.

Let A be the line given by the set of all (x, y) with $y = 1$. Is A bounded? Is it true that, for some point p and positive number c, all points of A are within c of p? Let's try a choice, say p the origin, and $c = 5$. But this doesn't work, for it isn't true that all points of A are within 5 of the origin. Clearly, no p and c will work. We conclude, then that this set A is not bounded. [This little discussion isn't quite a

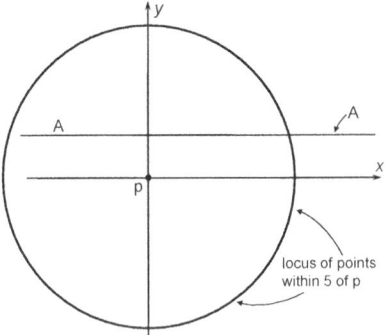

full proof of our conclusion. It would not be difficult to give a full proof at this point, but we shall not do so because the result isn't worth the effort. We shall discuss proofs in a moment.]

Finally, let A be the set of all (x, y) with $y = 1$ and x an integer. This set is a "bunch of equally spaced points", as shown. This set, clearly, is not bounded.

We turn, finally, to our last intuitive notion, that of "connected". [In order to avoid having sentences with many "nots" in them, let us deal with its opposite, "disconnected".] Recall that we wish to regard a set in the plane as disconnected if it "consists of two or more pieces, not joined together". Again, we wish to make this idea precise.

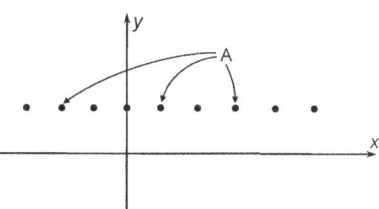

Consider, as an example, the set A consisting of two disks in the plane, as shown. What is it about this set which suggests to us that it consists of "two pieces"? One might say that the essential feature is how A is expressed, in this case, "A is the set of all (x, y) with either

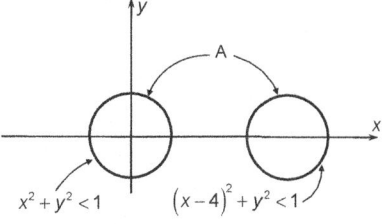

$x^2 + y^2 < 1$ or $(x - 4)^2 + y^2 < 1$". Note the use of "either ... or ...". Is this the tipoff that the set is disconnected? It is a good clue, but this would never serve as a mathematical definition of "disconnected". The problem is that it refers to *how the set is expressed* (in English), and not to *what the set is* (i.e., which points are in and which out).

A more promising possibility might be based on the following observation. In the case of our two disks, one can find a curve (such as that shown) such that part of the set is on one side of the curve, part is on the other side, and none of the set meets the curve. This curve, then, makes explicit the feature that the set "consists' of two or more

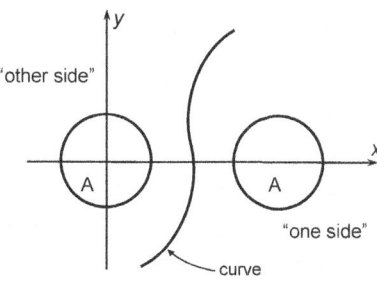

pieces". A definition along these lines, however, involves various other complications. What are we to mean by a "curve" in the plane? What is "one side" of the curve, and the "other side" to mean? That one has a good intuitive idea of what these words mean is not enough, for after all, one also has a good intuitive idea of what "disconnected" means. Perhaps one could resolve these difficulties as follows. Recall, from plane geometry, that, given two points in the plane, the perpendicular bisector of the segment joining those points, the line indicated in the figure, is just the locus of points equidistant from the two points. The points on one side of that perpendicular bisector are nearer to p than to p'; those on the other side nearer to p' than to p.

In this way, one might formulate the following: A set A is disconnected if there exit points p and p' such that some point of A is closer (using our distance) to p than to p', some point of A is closer to p' than to p, and no point of A is equidistant from p and p'. This works, for example, for our set A consisting of two disks. Unfortunately, this is also not a very good definition. The problem now is that it doesn't give the "intuitive answer" for certain examples.

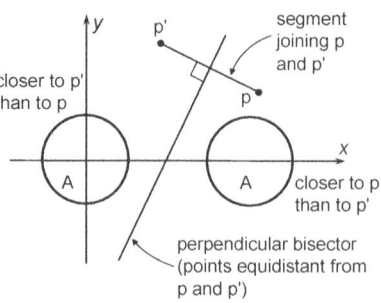

Consider, for instance, the set A consisting of "horseshoe and a disk" as shown. [We do not bother with writing the equations.] We would presumably wish to regard this set as disconnected. But it would not be recognized as disconnected according to the trial definition above. The problem here is that, although this set clearly seems to "consist of two pieces", and although one can, intuitively "separate these pieces by means of a curve", one cannot "separate those pieces by means of a straight line, i.e., by means of a perpendicular bisector". [No straight line can "wander inside the horseshoe to have the disk on the other side of the line from the horseshoe".] The problem with this definition, then, is that it doesn't properly reflect our intuition.

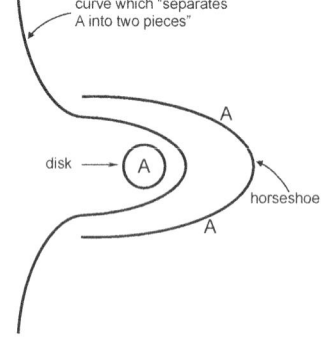

The key to what will ultimately be the definition of "disconnected" is to ask if we already have in our repertoire of definitions anything which is rather "curve-like". The answer is that we have: boundaries tend, in a very rough and general way, to be "curve-like things which separate the interior of a set from the exterior of that set". [Recall, for example, the situation for the disk.] Thus, we do not really need to find fancy definitions of "curve", "one side of the curve" and "the other side of the curve"'. We can use as surrogates "the boundary of a set", "the set itself" (which one may think of as being "on one side of the boundary"), and "the points outside of that set" (which one thinks of as being "on the other side of the boundary"). These observations, then, suggest that we try to capture the essential intuitive content of "there exists a curve such that part of the set is on one side, part on the other, and none of the set meets the curve" as follows.

<u>Definition</u>. Set A in the plane will be said to be *disconnected* if there exists

a set B in the plane such that some point of A is in B, some point of A is not in B, and no point of A is in bnd (B).

Here, if I have ever seen one, is an example of a good idea in mathematics. One combines the intuition we have of "boundary", the fact that we have formulated a definition of "boundary", and what we want to say by "disconnected" to obtain a candidate for a precise definition of "disconnected". To come up with original ideas, such as this one, for a definition is an activity to which many mathematicians devote a great deal of effort. [This definition is an old one. It is not, of course, original with me.] Of course, we do not wet know whether or not this definition will turn out to be an appropriate one, i.e., whether it will give the expected answer in simple examples, but it certainly looks promising, What remains, then, is to simply try it out.

Let us return, then, to our example of two disks. Is this set, according to our definition, disconnected? Can we find a set B such that some point of A is in B, some point is not in B, and no point of A is in bnd (B)? It turns out that we can. [To find such a set B, one thinks about how we arrived at the definition. One, in one's mind's eye, 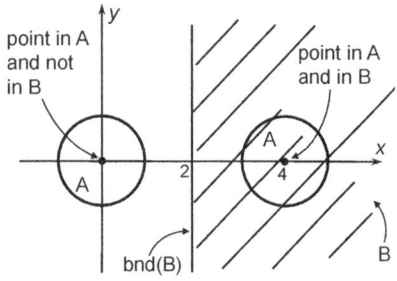 draws "a curve between the two pieces of A". Then one takes "one side of that curve", and designates that set "B".] Let, then, B be the set of all points (x, y) with $x > 2$ – the set illustrated in the figure. Then certainly some point of A is in B (namely, for example, the center of the "right" disk the point $(4, 0)$) and some point of A is not in B (namely, "left" disk, the point $(0, 0)$). Further, bnd B is just the line indicated, i.e., the set of all points (x, y) with $x = 2$. [One, of course, must check with the definition of "boundary" to get this.] Clearly, no point of bnd (B) is in A. So, this A satisfies the conditions of the definition (for we have found a set B such that some point of A is in B, some point of A is not in B, and no point of A is in bnd (B). So, this set A is disconnected. Similarly, a set A consisting of two points is disconnected.

Next, consider the set A consisting of the "horseshoe and disk", as on page 20. This set is also disconnected. Let B, for example, be the set shown. [Again, we will not bother to write down equations.] Then, again, some point of A is in B, some point of A is not in B, and no point of A is in bnd (B). So, this set is disconnected.

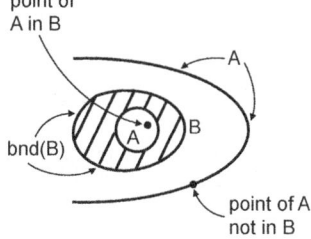

Our "two squares" set A, as illustrated on page 14, is also disconnected. Now, let B consist of all points (x, y) with $x > 1$. Then bnd (B) is all (x, y) with $x = 1$. Now, there is a point (such as q in the figure) in A and in B, a point (such as q') in A and not in B, while no point of A is in bnd (B). [For the latter, the reason is that the "line between the two squares" through which

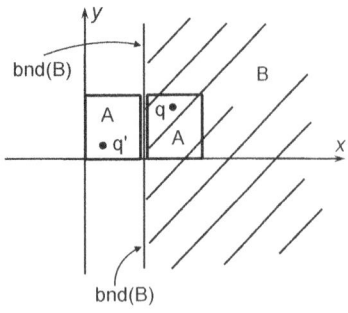

"the boundary of B passes" is not included in our set A.] So, this B will do the job. This set A is disconnected.

We give two examples of sets which are connected (= not disconnected). Let A be the set consisting of the single point at the origin. Does there exist a set B such that some point of A is in B, some point of A is not in B, and no point of A is in bnd (B)? Hardly, for A only has one point, and this point would have to be either in B or not in B, no matter what B is. There

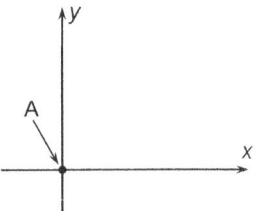

could hardly be *both* a point of A in B and a point of A not in B, since A has only one point.

Now consider, as a second example, a line, e.g., the set A consisting of all (x, y) with $y = 1$. This set, we claim, is connected. [A full proof of this assertion, while not terribly difficult, does involve some technical properties of the real numbers. Since we are here not particularly interested in the real numbers, we shall

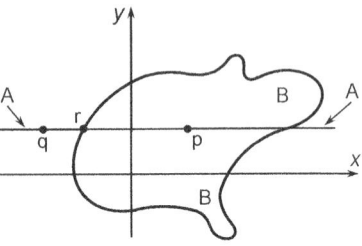

just give an informal argument.] Suppose, on the contrary, that one could find some set B such that some point of the line A is in B, some point of A is not in B, and no point of A is in bnd (B) (i.e., a set B which could cause A to be disconnected). We argue for a contradiction. Let p be the point of A in B, as shown in the figure. Then p cannot be in bnd (B) (for no point point of A is to be in bnd (B)). That is, it must be false that, for every positive ϵ, some point within ϵ of p is in B and some within ϵ of p is outside of B. But certainly for every positive ϵ, some point within ϵ of p is in B, namely p itself (clearly within ϵ of p and clearly in B). So, the "other part" must fail. That is, it must be false that, for every positive ϵ, some point within ϵ of p is outside of B. That is, there must be some positive ϵ such that *every* point within ϵ of p is in B. But this just means that the point p is in int (B).

Next, let q be the point of A not in B, as shown in the figure. Then this q is certainly *not* in int (B), for q is not even in the set B itself. Now start at the point p (which is in int (B)) and work toward point q (which is not in int (B)), and continually ask, as one moves along, whether one is still in int (B). Eventually, the answer will have to change from "yes" to "no". Let r be the point at which this change takes place. [The existence of such an r requires the "technical property of the real numbers".] This r, we now claim, must be in bnd (B). Indeed, for every positive ϵ, there is a point within ϵ of r and in B (since the points "just to right of r along the line" are of course in B). Furthermore, for every positive ϵ, there is a point within ϵ of r and not in B (for should this fail, i.e., were it the case that for some positive ϵ, every point within ϵ of r is in B, then r would be in int (B). But r was chosen to be the first point which ceases to be in int (B).) Putting these two together, we conclude that r is in bnd (B). Now we have our contradiction, for the point r is certainly in A (i.e., on our line) and is also in bnd (B), while our B was supposed to have the property that no point of A is in bnd (B).

The above is, admittedly (and unfortunately), a rather complicated argument, but I hope that its sense is clear. In order that the line A be disconnected, one must find set B such that the line gets from a point (p) in B to a point (q) not in B, all without encountering any point of bnd (B). But this cannot happen, because a transition must somewhere (on the line) take place, yielding a point of the line in bnd (B).

In a similar way, one argues that the disk is disconnected. Suppose that one could find a set B such that some point of A is in B, some point of A is not in B, and no point of A is in bnd B, as shown. The line segment joining p (in A and in B) and q (in A and not in B) will lie entirely within our disk A. But this line segment goes from a point p in B to a point q not in B, and so, arguing as before, we find a point r of this line segment in bnd (B). We thus obtain our

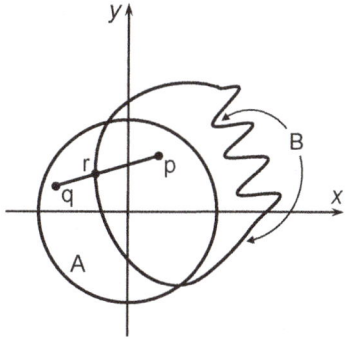

contradiction, for this r is in A (since r is on our line segment, and the whole line segment is in A), and also in bnd (B).

Similarly the entire plane and the empty set are both connected.

As an example of all four definitions, we consider again the set A of the second figure on page 4: the set of all (x, y) with x rational. In this case, bnd (A) will be the entire plane (for no matter what point p and what positive ϵ one chooses, there will be some point within ϵ of p and in A, and also some point within ϵ of p and not in A. Int (A) will be the empty set (for, for no point p can one find a positive ϵ such that every point within ϵ of p is in our set A). This A will not be bounded (for, for no point p and positive number c will every point of A be within c of p), and it will be disconnected (for, choosing, say, B the set of all points (x, y) with $x > \pi$, there will certainly be a point of A in B and also a point of A not in B, while no point of A is in bnd (B), for bnd (B) is the set of all points (x, y) with $x = \pi$, while, since π is not rational, none of these points will be in A).

We shall make extensive use of these, as well as other, similar, examples. The skills which, it seems to me, are important here are the ability to apply the definitions mechanically in unfamiliar situation (which is not easy, because the definitions are filled with "for some positive ϵ", "every point of", "for some point", "and also", and so on), and the ability to recognize quickly what the answer is in more familiar situations.

4. Properties of Boundary, Interior, Bounded, and Connected

We began with certain intuitive notions about sets in the plane. By introducing the usual distance between points in the plane, we were able to "capture" these notions, i.e., to formulate mathematical definitions of "boundary", "interior", bounded" and "connected'. The next step was to test these definitions and understand what they are saying. In particular, one wants to know whether or not the definitions faithfully reproduce our own ideas about what "boundary", "interior", "bounded" and "connected" should mean. This "testing and understanding" is normal operating procedure following the introduction of any definition in mathematics. The procedure usually consists of two stages. In the first stage, one applies the definition to various examples (preferably, to examples as diverse as possible), in order to get a feeling for the scope of the definition and how it works. We have now completed this stage. In the second stage, one looks around for properties of the definition: results, usually of an elementary sort, which involve the thing defined in some way, i.e., which relate it to itself or to other mathematical ideas with which one might be familiar. [The line between these two stages is in practice not as sharp as suggested here.] We now turn to this second stage. We emphasize at the outset that the properties of "boundary", "interior", "bounded", and "connected" number well into the hundreds, and probably into the thousands. It is thus not practical to try to list *all* of the properties: nor is it practical to try to memorize those properties one does find. Rather, one wishes to obtain an overall feeling for what is true and what is false, what argument is likely to yield a proof of an assertion and what argument is not, what examples are likely to show that an assertion is false and what examples are not.

Is there any simple relation between a set A and its interior? Let us try a few examples. The interior of the entire plane is the entire plane; the interior of the disk is the disk; the interior of a line is the empty set; the interior of a point is the empty set; the interior of the empty set is the empty set. Can one see any pattern to these examples? Thinking about these for a while, and possibly trying a few more, one notices that there is indeed a pattern: the interior of a set seems never to be "larger" than the original set. So, we have now discovered what seems to be a pattern in various examples. We next wish to see if we can formulate some precise statement which reflects the pattern we have found. [The statement above is unsuitable, for we have no defined "larger".] In this example, it is easy to say what one really means, namely that int (A) never "gets outside" of A, i.e., that every point of int(A) is also in A. Finally, we face the task of deciding whether this pattern is just a coincidence or a special property of our particular examples, or whether it will hold for *all* examples. What does it means for a point p to be in int (A)? It means that, for some positive ϵ, *every* point within distance ϵ of p is in A. But what about p itself? It is certainly within ϵ of p (for, in fact, $d(p, p) = 0$). So, it would have to be in A too. Clearly, then, it will be true for *any* example (i.e., for any set A) that every point of int (A) is in A. So, we have found a pattern, expressed it, and decided that it will actually hold for any example. The final step is to express all this in the "theorem-proof" format of mathematics. For the present case, a suitable rendition is:

Theorem. Let A be a set in the plane. Then every point of int (A) is in A.

Proof: Let point p be in int (A). Then, for some positive number ϵ, every point within distance ϵ of p is in A. But p is within distance ϵ of p, and so p must be in A.

The writing of theorems and proofs in mathematics is very much an art form. As such, it is a skill which is learned more through exposure to numerous examples than trough following rules. We shall, of course, see many examples of theorems and proofs. A few general guidelines at this point may, however, be useful. Consider first the words and symbols which follow "theorem". These should be full sentences (in which symbols, of course, are allowed). These words must represent a clear and unambiguous assertion that something is true. Thus, for example, "Let A be a set in the plane." would not be appropriate, for it does not assert anything, while "Let A be a disconnected set in the plane. Then A consists of two or more pieces." would not be appropriate either, for, while it certainly seems to be trying to assert something, it is not clear exactly what (for we do not know what "consists of two or more pieces" means).

One does not ordinarily omit even obvious conditions in the statement of a theorem. Thus, for the above, "Every point of int (A) is in A.": What is A? a point? a number? [One is in practice a bit more flexible than this. Completely obvious conditions are sometimes omitted.] It is, however, always incorrect to omit significant conditions. If, for example, the theorem above were true only when the set A is bounded, then the omission of this condition (even if it were made clear in the proof that A has to be bounded, and even if the preceding discussion was all about bounded sets) would be inappropriate. Finally, there should be nothing in the statement of the theorem which is irrelevant to the assertion it makes. The following, for example, would not be good theorems: "Let A and B be sets in the plane. Then every point of int (A) is in A."; "Let A be a set in the plane. Then every point of int (A), which represents all points which are 'well inside' A, is in A." A proof should, of course, consist of sentences, and represent a clear demonstration that the assertion of the theorem is in fact true. The proof should be the most direct line one can find, and should of course contain nothing irrelevant to what is being demonstrated. Thus, for example, replacing the first sentence by "Last point p be in int (A), so p is 'well inside' A", replacing the last sentence by "What about the point p itself? This p is within distance ϵ of p and so p must be in A." or adding a sentence at the end such as "For, were p not in A, then it could not be within ϵ of p, and that would give a contradiction, since p is within 0 of p, and so is within ϵ of p." would not be appropriate. Every sentence of the proof should be clear: It should be clear what one is using, what one is concluding, and why. Thus, "Since for some positive every point is within ϵ of p is in A, p must be in A". is just not clear. [What is this p? Are we assuming that p is in int (A)?] The various steps in the proof should be approximately equal in difficulty. That is, instead of five very simple steps followed by one very difficult one, one would combine some of the five initial steps and break up the last one. The difficulties of the successive steps should be such that one can see clearly how to get from one to the next. The following proof, for example, contains too many steps. "To show that every point of int (A) is in A, it suffices to select one point, p, in int (A), assuming nothing more about p, and show that this p is in A. But since p is assumed to be in int (A), it must satisfy the definition.

That is, for some positive ϵ, every point within ϵ of p must be in A. But, no matter what the positive ϵ is, the point p itself is within ϵ of p, for in fact the distance of p from p is zero. So, since every point within ϵ of p must be in A, and since p is within ϵ of p, the point p must be in A. So, since we began with any point p in int (A), and showed that it is in A, we have shown that every point in int (A) is in A." Finally, one does not normally let any motivation or other thoughts one might have creep into the proof.

Is there any simple relation between set A and its boundary? Is, for example, every point of bnd (A) in A, or every point of A in bnd(A)? Let A be the disk (all (x, y) with $x^2 + y^2 < 1$), so bnd (A) consists of all (x, y) with $x^2 + y^2 = 1$. In this example, there are points of A not in bnd (A), and points in bnd (A) not in A. This example, however, suggests that no point can be in both bnd (A) and A. Is this true? Unfortunately, it is not in general: For A a line, bnd $(A) = A$, and so A and its boundary have points in common. There is, apparently, no simple relationship between A and its boundary.

Since the boundary or interior of a set is again a set, one can take the boundary or interior again. Are there any simple relations? Consider int (int (A)), i.e., the set obtained by first determining the interior of A, then the interior of the resulting set. We try some examples. For A the entire plane, int (A) is the entire plane, and so int (int (A)) is the entire plane; for A the disk, int (A) is the same disk, and so int (int (A)) is the disk; for A a line, or a point, or the empty set, int(A) is the empty set, and so int (int (A)) is also the empty set.

After perhaps trying a few more examples, one realizes that there is a pattern here: What one gets at the second step seems always to be just what one gets at the end. We are led, then, to guess that, for any set A, int (int (A)) = int (A). Is this true in general? Suppose, first, that we have a point of int (A)). Then, by the Theorem on page 26 applied to "int (A)", it follows that our point must be in int (A). So, we have only to show the reverse.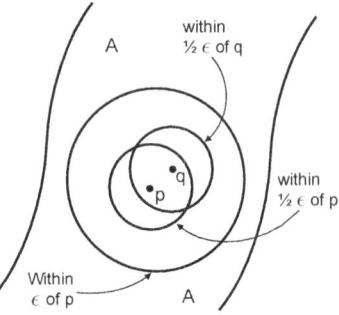
The idea is the following. Suppose p is a point of int (A). Then all points within some ϵ of p are in A, as shown in the figure. Now consider the disk centered at p, with radius $1/2\,\epsilon$. All of the points in this disk are of course in A. But all of these points must also be in int (A), for clearly, since all points of the larger disk are in A, all points of the smaller disk must be in int (A). The points of A "fill" the larger disk, and so A includes the smaller disk in its interior. We see, then, that our guess is in fact true. All this might be expressed as follows.

Theorem. For any set A in the plane, int (int (A)) = int (A).
Proof: By the Theorem on page 26, every point of int (int (A)) is in int (A). For the converse, let p be any point of int (A). Then for some positive ϵ, every point within ϵ of p is in A. Let q be any point within $1/2\,\epsilon$ of p. Then every point within $1/2\,\epsilon$ of q must be within ϵ of p, and so must be in A. So, q must be in int (A) Since every point within $1/2\,\epsilon$ of p is in int (A), p is in int (A), p is in int (int (A)).

Are there other such relations? What about int (bnd (A))? For the entire plane, bnd (A) is the empty set, so int (bnd (A)) is the empty set: for A the disk, bnd (A) is a circle, so int (bnd (A)) is the empty set; for A a line, or a point, or the empty set, bnd (A) is just A, so int (bnd (A)) is the empty set. The pattern is obvious: We always get the empty set. So, one might guess that always int (bnd (A)) is the empty set. One might at this point try for a while to see if a proof could be found – but one would fail. In fact, this is false. Consider, for instance, the example at the bottom of page 23: A is all (x, y) with x rational. Then bnd (A) is the entire plane, and so int (bnd (A)) is the entire plane! There is no simpler expression for int (bnd (A)).

For S any set in the plane, the *complement* of A, denoted A^C, is the set of all points not in A. [Note that, given the unambiguous statement of what is in and what is not in A, we obtain an unambiguous statement of what is in and what is not in A^C.] For example, the complement of the entire plane is the empty set; the complement of the disk is the set of all (x, y) with $x^2 + y^2 \geq 1$ (the plane with "a hole in it"). The complement of the complement of A is again the set A.

What relations are there between the boundary of interior of A^C and A? Consider int (A^C). For A the disk, A^C is the set of all (x, y) with $x^2 + y^2 > 1$, and so int (A^C) is the set of all (x, y) with $x^2 + y^2 > 1$. For A consisting of the single point $(0, 0)$, A^C consists of all points except $(0, 0)$, and so int $(A^C) = A^C$. For A the entire plane, A^C is the empty set, and so int (A^C) is the empty set. There is in fact a pattern here, namely that no point is in both A and int (A^C). This is true in general. Indeed, were a point in both A and int (A^C), then, by the Theorem on page 26, it would be in both A and A^C, which contradicts what "complement" means.

A more interesting relation involves the boundary. Imagine taking a photograph of the set A, with A itself in white and the background (A^C) in black. Then the negative of this photograph (with A^C in white and A in black) would be the photograph of A^C. Now think of bnd (A) as the "border" between the black and white regions of the original photograph. Then one might expect that this would be the same as the border between the white and black regions of the negative. These intuitive remarks suggest:

Theorem. Let A be a set in the plane. Then bnd (A^C) = bnd (A).

Proof: Point p is in bnd (A) if and only if, for every positive number ϵ, there is some point within ϵ of p and in A, and also some point within ϵ of p and not in A. But, since the points in A are precisely those not in A^C, this holds if and only if, for every positive number ϵ, there is some point within ϵ of p and not in A^C, and also some point within ϵ of p and in A^C, But this holds, finally, if and only if the point p is in bnd (A^C).

[In mathematics, "... if and only if ···" means "whenever ··· holds, ... also holds, and furthermore whenever ... hods, ··· also holds."] Our previous two theorems were rather ugly things, designed primarily just to illustrate how one discovers theorems, how one states them, and how one proves them. This theorem, however, is a bit more interesting. Its nice feature is that one guesses it from one's ideas about "boundary". It genuinely seems to reflect an aspect of what a "boundary" should be. It furthermore, with our definition of "boundary" turns out to be true. One's confidence in the appropriateness of the definition is strengthened.

For A and B sets in the plane, we say that A is a *subset* of B, written $A \subset B$, if every point in A is also in B.

For example, every set in the plane is a subset of the entire plane; the set consisting only of $(0, 0)$ is a subset of the disk; the empty set is a subset of any set in the plane. If A is a subset of B and B a subset of C, then A is a subset of C.

There are apparently no simple relations connecting subsets and boundaries. [For example, it is not true that, if $A \subset B$, then bnd $(A) \subset$ bnd (B). Let A consist of the single point $(0, 0)$ and B the usual disk. Then $A \subset B$. But bnd $(A) = A$, and bnd (B) consists of all (x, y) with $x^2 + y^2 = 1$, so it is not true that bnd $(A) \subset$ bnd (B).] For the interior, however, we have

Theorem. Let A and B be sets in the plane, with $A \subset B$. Then int$(A) \subset$ int (B).

Proof: Let p be a point of int (A). Then, for some positive number ϵ, every point within distance ϵ of p is in A. But, since $A \subset B$, every point within distance ϵ of p is in B. Hence, p is in int (B).

One interprets the theorem as saying "the smaller the set, the smaller its interior".

For A and B sets in the plane, the *intersection* of A and B, denoted $A \cap B$, is the set of all points which are in both A and B. Thus, for example, the intersection of the usual disk and the line given by all (x, y) with $y = 0$ is a "segment of a line", given by all (x, y) with $-1 < x < 1$ and $y = 0$; the intersection

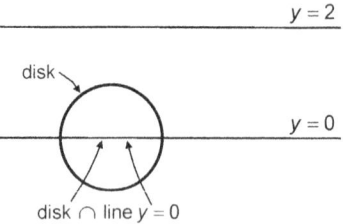

of this disk and the line given by all (x, y) with $y = 2$ is the empty set. The intersection of any set and the entire plane is that set; the intersection of any set and the empty set is the empty set. For any A and B, $A \cap B$ is a subset of A.

We give one example of a property involving intersections. Think of the points in the interior of a set as those "well inside" the set. Which points would one expect to be "well inside" $A \cap B$? Since $A \cap B$ consists just of those points which are in both A and B, one might expect that the points "well inside" $A \cap B$ will be just those which are both "well inside" A and "well inside" B. One tries a few examples. (For instance, for A the entire plane and B ant set, $A \cap B = B$, and so int $(A \cap B) =$ int (B); but int (A) is the entire plane, so int $(A) \cap$ int $(B) =$ int (B). So, in this example, int $(A \cap B) =$ int $(A) \cap$ int (B).), and finds this to be the case. All this leads to

Theorem. Let A and B be sets in the plane. Then int $(A \cap B) =$ int $(A) \cap$ int (B).

Proof: Let p be a point of int $(A \cap B)$. Then, for some positive number ϵ, all points within ϵ of p are in $A \cap B$. It follows that all points within ϵ of p are in A, whence p is in int (A); and that all points within ϵ of p are in B, whence p in int (B). So, p is in int $(A) \cap$ int (B). For the converse, let q be a point of int $(A) \cap$ int (B). Then q is in int (A), and so, for some positive number ϵ_1, every point within distance ϵ_1 of q is in A; and q is in int (B), so, for some positive number ϵ_2, every point within distance ϵ_2 of q is in B. Let ϵ, a positive number, be the smaller of ϵ_1 and ϵ_2. Then every point within ϵ

of q is both in A and in B, and so is in $A \cap B$. That is, q is in int $(A \cap B)$.

This proof, like all the proofs we have had, consists merely of unravelling the definitions to figure out what one has, and then re-ravelling then to figure out what one gets. [Indeed, is some sense *all* proofs in mathematics are just this. Theorems in mathematics never have any

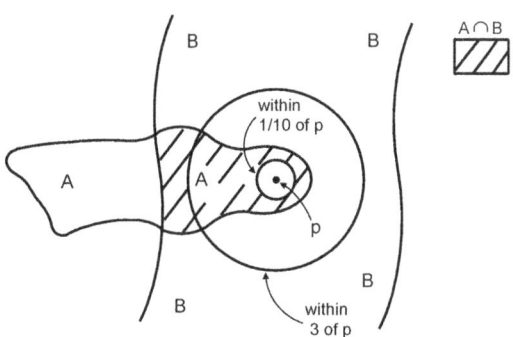

"real content".] This one, however, is a bit more complicated. for the second part, in particular, the idea is the following. One has point q in int $(A) \cap$ int (B), i.e., q is in both int (A) and int (B), and one wants to show that q is in int $(A \cap B)$. One has positive number ϵ_1, which specifies "how well inside" q is of A, and ϵ_2 for B. What should one choose for one's ϵ to describe "how well inside" q is of $A \cap B$? Suppose, for example, that ϵ_1 were $1/10$, and ϵ_2 were 3, as shown in the figure. That is, q is "just barely well inside" A, but "quite well inside" B. How "well inside" will q be of $A \cap B$? The answer is "just barely well inside", as one sees from the figure. These observations, then, suggest what should be chosen for the ϵ in the proof. [Of course, it is perfectly legitimate to draw a picture to aid in the discovery of a proof. But the picture should not be part of the actual proof.]

For A and B sets in the plane, the *union* of A and B, denoted $A \cup B$, is the set of all points which are either in A or in B. As examples, the union of two disks is the striped region illustrated at the right. The union of any set and the

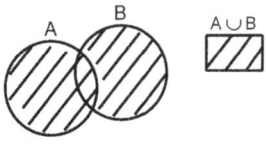

entire plane is the entire plane. The union of any set and the empty set is the original set. For any sets A and B in the plane, A, B, and $A \cap B$ are all subsets of $A \cup B$.

What relations are there between the boundary, interior, and union? From the theorem on the previous page, one might guess that perhaps int $(A \cup B)$ = int $(A) \cup$ int (B). Indeed, this seems somewhat reasonable intuitively, for the points "well inside" $A \cup B$ ought to be those either "well inside" A or "well inside" B. Unfortunately, this is false. Let A be the set of all points (x, y) with x rational, and B the set of all points (x, y) with x irrational (= not rational). Then int (A) and int (B) are both the empty set. Hence, int $(A) \cup$ int (B) is the empty set. But $A \cup B$ is the entire plane (for *every* point (x, y) in the plane has either x rational or x irrational, and so will be in either A or B. In fact, $A^C = B$.) Hence, int $(A \cup B)$ = int (entire plane) = entire plane. So, in this example, we do not have int $(A \cup B) = $ int $(A) \cup$ int (B). In retrospect, we can see what was wrong with our intuition. It is possible, as in this example, to have a point "well inside" $A \cup B$ because A and B work together to "overrun" the vicinity of the point. However, each of A and B individually contains "gaps very near the point", so that the point turns out not to be "well inside", either A or B by itself. Perhaps, however, we can save the other half of our guess. Perhaps every point of int $(A) \cup$ int (B) is in int $(A \cup B)$. This works.

Theorem. Let A and B be sets in the plane. Then int $(A) \cup$ int $(B) \subset$ int $(A \cup B)$.

Proof: Let p be a point in int $(A) \cup$ int (B), say p in int (A). Then, for some positive number ϵ, every point within distance ϵ of p is in A. But now every point distance ϵ of p is in $A \cup B$, and so p is in int $(A \cup B)$.

[Note that in this proof we omitted "Since p is in int $(A) \cup$ int (B), p is in either int (A) or int (B). We shall do the case p in int (A); that for p in int (B) is done similarly."]

We have dealt so far only with properties of the interior and the boundary. We next consider briefly a few properties of bounded and connected.

Is the complement of a bounded set bounded? Certainly not: A disk is bounded but its complement is not. Indeed, quite the opposite is true: The complement of a bounded set can never be bounded. If $A \subset B$ and A is bounded, need B be? No: let A be the disk and B the entire

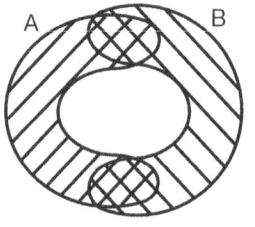

plane. In fact, this is the wrong way around. "Smaller sets" are more likely to be bounded.

Theorem. Let A and B be sets in the plane, with $A \subset B$. Then, if B is bounded, so is A.

Proof: Let B be bounded, and let p be a point and c a positive number such that every point of B is within distance c of p. Then, since $A \subset B$, every point of A is within distance c of p, and so A is bounded.

It follows immediately that the intersection of two bounded sets is bounded (since that intersection is of course a subset of one of the sets). Is the union of two bounded sets bounded?

Is the complement of a connected set connected? No: Let the set be a straight line. Then its compliment is "everything except the line", i.e., two half-planes, and this set is not connected. Is the subset of a connected set connected? The answer is again no. A set consisting of two points is a subset of the entire plane, the entire plane is connected, but the set consisting of two points is not. Is the intersection of two connected sets connected? The answer is again no, but now the example is a bit trickier. The sets illustrated in the figure will do the job. [Since it is clear how one would write the equations for and do in detail (i.e., using the definition) such an example, we will not do so here.] Finally, the union of two connected sets need not be connected: Let each of A and B be a disk of unite radius, but let them be "well separated in the plane". Then each is of course connected, but their union, "two well-separated disks", is not connected.

Does "connected", then, have no simple properties? It turns out that it does have one very nice property. Let us consider again the final example above: the union of two disks. Why, in that example, was $A \cup B$ disconnected? The reason, of course, was that "$A \cup B$ consisted of two pieces, namely A and B separately". It would seem intuitively that this might be the only situation under which A and B are themselves connected while $A \cup B$ is disconnected. Can we make this general idea into a theorem? How do we state "for A and B connected, $A \cup B$ is either connected, or consists just of the two pieces A and B". One possibility, recalling the definition and interpretation of "connected", would be: "Let A and B be connected, and let C be a set such that some point of $A \cup B$ is in C, some point of $A \cup B$ is not in C, and no point of $A \cup B$ is in bnd (C). Then either $A \subset C$ and $B \subset C^C$, or $A \subset C^C$ and $B \subset C$". [This is the right general idea, for, recall, C and C^C are to be the "two pieces" into which C divides the plane.] In fact, this statement is true. It turns out, however, that there is a simpler way to state the same idea. We simply *demand* that A and B be not "completely separate from each other" (so they cannot be the "two pieces" of $A \cup B$) by demanding that they have a point in common.

Theorem. Let A and B be connected sets in the plane, and let p be a point of both A and B. Then $A \cup B$ is connected.

Proof: Suppose, for contradiction, that C were a set such that some point of $A \cup B$ (say, u in A) is in C, some point of $A \cup B$ (say, v in B) is not in C, and no point of $A \cup B$ is in bnd (C). Then no point of A and no point of B is in bnd (C). But point u in A is in C and no point of A is in bnd (C), and so, since A is connected, every point of A must be in C. In particular, point p must be in C. But now point p in B is in C, point v in B is not in C, and no point of B is in bnd (C), which contradicts the fact that B is connected.

I hope that the structure of the proof is clear. We suppose that $A \cup B$ could be "separated into two pieces" by means of set C. In order for this to be a "genuine separation", some point (u in A) must be in C, and some point (v in B) must be out of C. But A itself cannot be "separated into two pieces" by C, because A is assumed connected. So, A must be either all in C or all out of C. But A has the point u in common with C, and so A must end up "all in" C. But now, having dragged all of A into C, we have also brought the point p (common to both A and B) into C. Now C goes to work on B. Since a little bit of B (namely, the point p) has been dragged into C, and since B is also connected, all of B must be in C too. Here, then, is a contradiction, since the point v of B was supposed to be not in C. [Of course, one could put

u and v in other places in $A \cup B$, with the same result.]

This is a nice theorem. It expresses very well an intuitive connotation of "connected", and it is even true.

We emphasize that the purpose of this section has been to illustrate what one does to "understand" a definition in mathematics. The process above is what a mathematician would do automatically on being confronted with a new definition. Of course, the various theorems and examples above are only to illustrate a few of the many possibilities: They are not to be memorized. A second purpose was to illustrate the type of thought-process one goes through in inventing theorems and inventing proofs. Finally, the stated theorems and proofs are intended to be examples of that "art form".

5. Continuous Curves

We conclude our discussion of the plane with a final set of definitions. These are perhaps a bit more complicated than those we have treated in the previous sections. Here, we shall merely motivate the definitions, state them, and indicate what they mean. We shall return to this subject later.

One has an intuitive idea of what a "curve" in the plane ought to be: It is a "line", or "what results if one presses a piece of a chalk against a blackboard, and then moves the chalk about". [In mathematics, "curve" is always a generic term, which does not exclude "straightness". Thus, a straight line is also to be a "curve".] Our first goal is to formulate a definition of "curve".

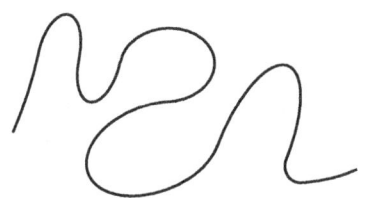

Let us begin by looking for a definition of the form "A set A in the plane is a curve provided ... " The "..." is supposed to express the idea that the set A must be "one-dimensional". How shall we say this idea? Recall that, for A a straight line ("one-dimensional"), int (A) is the empty set, while for A a disk ("not one-dimensional") int A is not the empty set. Thus, one might be tempted to replace "..." by "int (A) is the empty set". This, however, is not a very good choice. For example, let A be the set of all points (x, y) with x rational. Then int (A) is indeed the empty set, but we would certainly not wish to call this A a "curve". Alternatively, recalling that boundaries tend to be "one-dimensional", one might replace "..." by "for some set B, $A =$ bnd (B)". But this is not too good either, for, for example, the entire plane is the boundary of *some* set B (namely, B all (x, y) with x rational), while we would certainly not wish to regard the entire plane as a "curve". A few more failures along these lines convince one that this is not a very good task. [Of course, one possibility at this point would be that there is not going to be any "good" definition of a curve. But in this case, as it turns out, there will be.]

The key to getting a definition of "curve" is to think again about the chalk on the blackboard. In the attempt above, we were trying to characterize a curve in terms of just the set A of points of the plane "visited by the chalk".

But intuitively a curve would seem to be much more than merely "the set of all points visited".

There is also involved, for example, the order in which theses points are visited as the chalk moves over the blackboard. Can we somehow find a definition which somehow brings to mind the actual physical process of moving chalk over blackboard? We can. Imagine that a clock were running as the curve is being drawn on the blackboard. Then at any time t, as read by the clock, the chalk will be at some position on the blackboard. At various other times, the chalk will be at various other positions on the blackboard. Thus, as "time" runs through its various values, "position of the chalk" will run through various points of our plane. The idea, then, would be to regard a "curve" as, not merely the set of points of the plane "visited", but also as the "labelling" of those points by a parameter t, the "time". These ideas suggest.

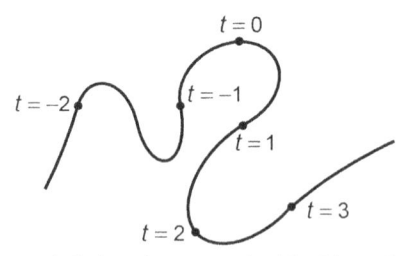

Definition. A *curve* in the plane is a rule which assigns unambiguously, to each real number t, a point of the plane.

We shall represent a curve symbolically by a letter such as γ. Then, for γ a curve and t a real number, we denote by $\gamma(t)$ that point of the plane assigned to the number t by the unambiguous rule which is the curve. Of course, one thinks of this t as the "time" as recorded by the clock, and of the point $\gamma(t)$ of the plane as the "location of the chalk on the blackboard at time t".

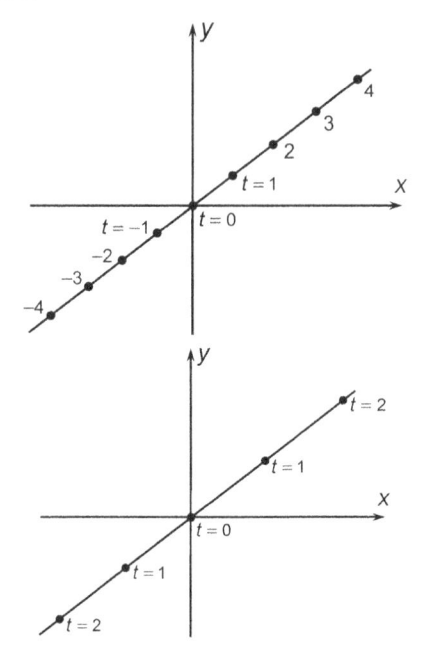

We give some examples of curves. Let γ be the rule which assigns, to the real number t, the point of the plane with $x = t$ and $y = t$, i.e., the point with coordinates (t, t). This is certainly an unambiguous rule: For example, it assigns to the number 93 the points of the plane with coordinates $(93, 93)$. This curve is illustrated in the second figure at the right.

It corresponds physically to "moving the chalk at uniform speed along a straight line tilted 45^0 from the vertical". Next, consider the rule which assigns to number t the point $(2t, 2t)$ of the plane. Again, this is an unambiguous rule: To the number -7, for example, it assigns the point $(-14, -14)$ of the plane. This curve is that shown. Note that this curve describes the same set in the plane as our previous curve. It is, however, to be regarded as a different curve, because we have a different rule. [To the number $t = 3$, for example, the first curve assigns the point $(3, 3)$; the second, the point $(6, 6)$.] Physically, we are now "moving the chalk twice as fast along this line".

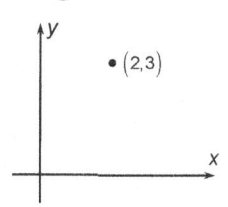

Let γ be the rule which assigns, to number t, the point $(2, 3)$ of the plane. This is unambiguous: It assigns, to $t = -137.9$, the point $(2, 3)$. This is thus a curve. it just "stay put" at the point $(2, 3)$. Consider the parabola $y = x^2$. This is *not* a curve according to our definition. It is not an unambiguous assignment of a point of the plane to each number t – and in fact it does not even say anything about t. What point does this "curve" assign to $t = 5$?

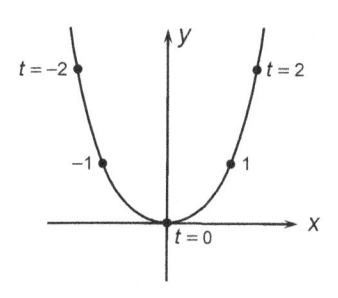

However, consider the rule which assigns, to number t, to point (t, t^2). [For example, to $t = 7$ is assigned the point $(7, 49)$. Clearly, this is "really" $y = x^2$.] Now, we do have a curve. In short, one does not specify a curve by giving "y as a function of x"; rather, one must give "each of x and y as a function of t". The rule which assigns to number t the point $(t^{13} - \cos t + t/(1 + t^2),$ $t^6 + t \sin t)$ of the plane is a curve. The rule which assigns to the number t the point of the plane which is $(2, 3)$ if t is a rational number, and $(9t, 61)$ if t is an irrational number is a curve. All we demand for a curve is an unambiguous rule (which, of course, assigns a unique point, $\gamma(t)$, of the plane to each number t) – nothing more and nothing less.

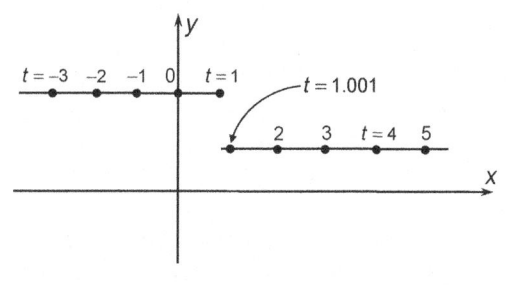

Some curves are better than others. Compare, for example, the curves illustrated in the figures on pages 36 and 37 with the curve of the last example above, or with the curve given by: For $t \leq 1$, let $\gamma(t)$ be the point $(t, 2)$, and for $t > 1$ let $\gamma(t)$ be the point $(t, 1)$ of the plane. This curve is illustrated on the right. [Note that the above is indeed an unambiguous rule to get from t's to points of the plane, hence, a curve. But "for $t \leq 1$, let $\gamma(t)$ be $(t, 2)$, and for $t \geq 1$ let $\gamma(t)$ be $(t, 1)$" is not a curve, for it is ambiguous. It assigns *two* points of the plane, namely $(1, 2)$ and $(1, 1)$, to $t = 1$.] The physical difference between the curves illustrated on pages 36 and 37 and the curve illustrated above is that the former can be "drawn without ever lifting the chalk from the blackboard", while the latter cannot. Our next task is to capture, by means of a definition, this distinction between curves which do and do not "skip".

To see what will be involved in making this definition, let us return to the curve (with a "skip") illustrated above. At what time did the skip take place? At $t = 1$. Clearly, we must isolate something unpleasant about what the curve was doing around $t = 1$. Where was the chalk at $t = 1$? At the point $\gamma(1)$, namely, the point $(1, 2)$, of the plane. What is it about the behaviour of the curve around $t = 1$ that leads us to regard the curve as having a "skip" there? The behaviour is the following. Whereas for $t = 1$ the chalk is at $\gamma(1) = (1, 2)$, for t just a tiny bit greater than 1, say $1 = 1.001$, the chalk is way over at $(1.001, 1)$. The problem, then, is that, during this "very short time" (from $t = 1$) to $t = 1.001$), the chalk has moved a "rather large amount" in the plane, namely from $\gamma(1) = (1, 2)$ to $\gamma(1.001) = (1.001, 1)$. Thus, to say that there is a skip at $t = 1$, we want to say that "when t moves just a little bit away from the value $t = 1$, the corresponding point, $\gamma(t)$ of the curve moves more than a little bit away from the point, $\gamma(1)$, of the curve corresponding to $t = 1$". All we have to do, then, is express these "little bits" precisely to obtain our definition. In the above example, the "amount of the skip", the "distance $\gamma(t)$ jumped between $t = 1$ and $t = 1.001$", was just a touch more than one unit. Thus, a statement which is true for this curve, and which reflects the fact that it has a skip at $t = 1$, is the following: There are t-values as close as you wish to $t = 1$ such that $d(\gamma(t), \gamma(0)) \geq 1/2$. [How close do you want the t-values to be to $t = 1$?. Within .01 of 1? Then $t = 1.001$ is this close. But $\gamma(1.001) = (1.001, 1)$, while $\gamma(1) = (1, 2)$, and so $d(\gamma(t), \gamma(1))$ is slightly more than one, and is certainly greater than or equal to $1/2$. Do you want the t-values still closer to 1?. You say you want them within .0000001 of $t = 1$? Then $t = 1.00000001$ is this close. But $\gamma(1.00000001, 1) = (1.00000001, 1)$, and its distance from $\gamma(1) = (1, 2)$ is still lightly more than one, and so is certainly greater than or equal to $1/2$.] A more precise version of "as close as you wish" in our statement above yields the following: Given any positive number δ, there is a t-value within δ of 1 (i.e., with $1 - \delta \leq t \leq 1 + \delta$) such that $d(\gamma(t), \gamma(1)) \geq 1/2$. The positive number δ represents of course "how close you wish t to be to 1." Instead of "t amount δ of 1", or "$1 - \delta \leq t \leq 1 + \delta$", we can write $|t - 1| \leq \delta$. Recall that $|\ |$ is the absolute value, the operation of reversing the sign if the number enclosed is negative and keeping the sign if it is positive, so, for example, $|6| = 6$, $|19| = 19$, while $|-3| = 3$ and $|-31| = 31$. So, $|t - 1| \leq \delta$ just says that "the number t is within amount δ of 1, i.e., is between $1 - \delta$ and $1 + \delta$.

In short, the "distance" between numbers a and b is just $|a - b|$.

So far, then, we have the following statement, which reflects the fact that there is a skip in the curve illustrated on the previous page: Given any positive number δ, there is a number t satisfying $|t - 1| \leq \delta$ and such that $d(\gamma(t), \gamma(1)) \geq 1/2$. A statement that there is "no skip" at $t = 1$ would thus be: There exists a positive number δ such that, whenever number t satisfies $|t - 1| \leq \delta$, $d(\gamma(t), \gamma(1)) \leq 1/2$. What this says is that " there exists *some* range of t-values around $t = 1$ (namely, the range of t-values consisting of those t between $1 - \delta$ and $1 + \delta$) such that, for all t in this range, the corresponding point $\gamma(t)$ of the plane is within distance $1/2$ of the point $\gamma(0)$". For example, consider again our curve at the end of page 37. Is it true that there is a range of t-values around $t = 1$ such that, for every such t, we have $d(\gamma(t), \gamma(1)) \leq 1/2$? It is certainly not true. The locus of points within distance $1/2$ of $\gamma(1)$ is the disk shown in the figure. there is no range of t-values around $t = 1$ such that for every t in this range $\gamma(t)$ lies within this disk. No matter what "range" (i.e., positive number δ) is given, one can find a t-value within this range (namely, a t-value very slightly greater than 1) such that $\gamma(t)$ is outside of this disk.

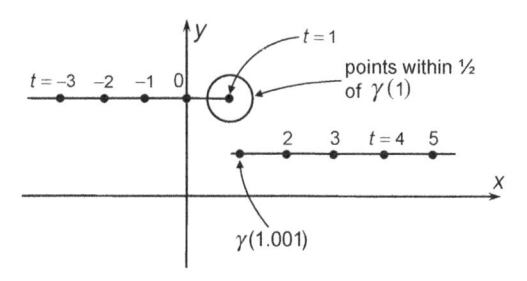

We now have practically the statement we are after. But where did this "1/2" come from? It arose originally from our example. We were looking at a curve with a skip of about 1, so we choose $1/2$ as the "amount of skip" to give ourselves a little leeway. What the statement above really says, then, is that there is no skip at $t = 1$ of "amount" greater than $1/2$. But we are not just interested in skips of amount greater than $1/2$: We want to rule out skips of *any* amount. We do so by: For *every* positive number ϵ, there exists a positive number δ such that, whenever t satisfies $|t - 1| \leq \delta$, $d(\gamma(t), \gamma(1)) \leq \epsilon$. This is to hold for *every* positive ϵ (the"amount of the skip"). It is to hold for $\epsilon = 1/2$ (no skips of amount greater than $1/2$), for $\epsilon = 1/10$ (no skips of amount greater than $1/10$), and so on. It thus says that there are no skips of any amount at $t = 1$. Finally, the discussion above has been for skips at the t-value $t = 1$. One can proceed similarly for other t-values. We thus arrive at:

Definition. Curve γ is said to be *continuous* at $t = t_0$ if, for every positive number ϵ, there exists a positive number δ such that, whenever $|t - t_0| \leq \delta$, $d(\gamma(t), \gamma(t_0)) \leq \epsilon$.

Thus, continuous at $t = t_0$ means "having no skips of any amount at $t = t_0$". For "having no skips of any amount anywhere", we have

Definition. Curve γ is said to be *continuous* if, for every number t_0, γ is continuous at $t = t_0$.

As usual in mathematics, one invests all one's thoughts and ideas in the *formulation* of a definition. However, once the definition has been stated, one essentially ignores all of one's ideas, and just mechanically follows the rules as laid down by that definition. Thus, in order to show that a curve is continuous at $t = t_0$, one has to figure out how, given any positive number ϵ, one can find a positive number δ such that, whenever $|t - t_0| \leq \delta, d(\gamma(t), \gamma(t_0)) \leq \epsilon$. We give some examples.

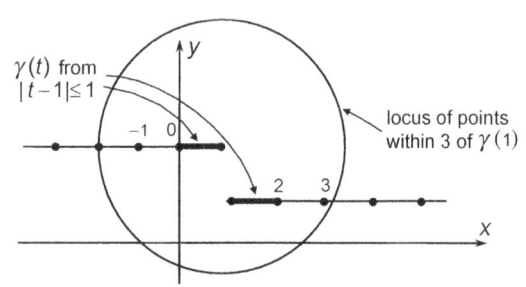

Consider the curve illustrated on the previous page. Is it continuous at $t = 1$? Is it true that, for every positive number ϵ, there exists a positive number δ such that, whenever $|t - 1| \leq \delta$, $d(\gamma(t), \gamma(1)) \leq \epsilon$? Let's try an ϵ, say $\epsilon = 3$. Is it true that there exists a positive number δ such that, whenever $|t - 1| \leq \delta$, $\gamma(t)$ is within 3 of $\gamma(1)$? The locus of points within 3 of $\gamma(1)$ is the disk shown. Does there exist a positive number δ such that, whenever $|t - 1| \leq \delta, \gamma(t)$ is in this disk? There certainly does: Choose $\delta = 1$. Then $|t - 1| \leq 1$ means that t is between 0 and 2, and so the corresponding $\gamma(t)$'s are those shown in the figure. But this part of the curve is certainly within our disk. So, there does exist a positive number δ such that, whenever $|t - 1| \leq \delta, d(\gamma(t), \gamma(1)) \leq 3$. So, we tried $\epsilon = 3$, and it worked. But we have to show this for *every* positive ϵ.

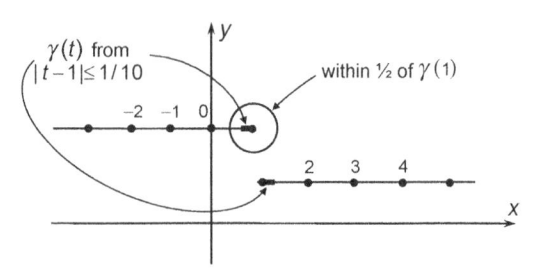

Let's try another, say $\epsilon = 1/2$. Is it true that there exists a positive number δ such that, whenever $|t - 1| \leq \delta, \gamma(t)$ is within $1/2$ of $\gamma(1)$? The locus of points within $1/2$ of $\gamma(1)$ is the disk shown. Does there exist a positive number δ such that, whenever $|t - 1| \leq \delta, \gamma(t)$ is in this disk? There does not. What value for δ will work? What about, say, $\delta = 1/10$. Is it true that, whenever t is between $9/10$ and $11/10$, $\gamma(t)$ is in this disk? It is not true, as one sees from the figure. Clearly, *no* δ will work. Thus, there does not exist a positive δ such that, whenever $|t - 1| \leq \delta, d(\gamma(t), \gamma(1)) \leq 1/2$. So, it is not true that, for *every* positive ϵ, there exists a positive δ such that, whenever $|t - 1| \leq \delta, d(\gamma(t), \gamma(1)) \leq \epsilon$ [for, as we have just seen, there is none for $\epsilon = 1/2$]. So, our curve is not continuous at $t = 1$. [This example gives the clue as to what one should choose for ϵ in order to show that a given curve is not continuous. Choose ϵ to be smaller than the "amount of skip".]

Is that curve continuous? Is it true that, for *every* number t_0, γ is continuous at $t = t_0$? It is not true, for this curve is not continuous at $t = 1$. Our definitions thus give the expected answers for this example.

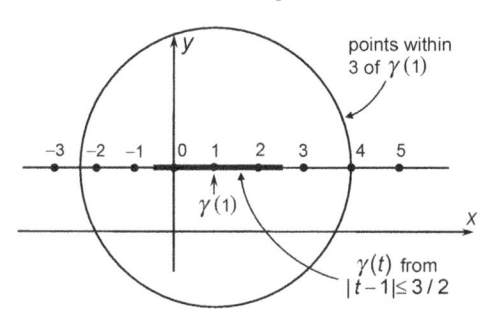

Consider, as a second example, the curve γ with $\gamma(t) = (t, 1)$. is this curve continuous at $t = 1$? Is it true that, for every positive ϵ, there exists a positive δ such that, whenever t is within δ of 1, $\gamma(t)$ is within ϵ of $\gamma(1)$? Let's try $\epsilon = 3$. Does there exist a positive δ such that, whenever t is within δ of 1, $\gamma(t)$ is within 3 of $\gamma(1)$? The locus of points within $3 of f \gamma(1)$ is the disk shown. We want to find a positive δ such that, whenever t is within δ of 1, $\gamma(t)$ is within 3 of $\gamma(1)$. But there is such a δ, namely for example. $\delta = 3/2$. The portion of the curve for $|t - 1| \leq 3/2$ is indicated in the figure, and all this portion is within our disk. So, we found a δ for this ϵ.

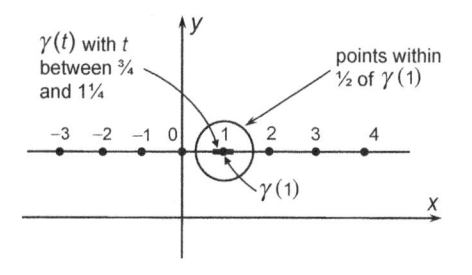

What about another, say $\epsilon = 1/2$? Does there exist a positive ϵ such that, whenever t is within δ of 1, $\gamma(t)$ is within $1/2$ of $\gamma(1)$? The locus of points within $1/2$ of $\gamma(1)$ is the disk shown. What do we choose for δ? Well, $\delta = 1/4$ will do, for, whenever t is within $1/4$ of 1 (i.e., between $3/4$ and $1\,1/4$), $\gamma(t)$ is within $1/2$ of $\gamma(1)$, as shown in the figure. Thus, we found a suitable δ for this choice of ϵ. One tries another, say $\epsilon = 1/10$. For this ϵ, one will have to choose a still smaller δ, for determines "how close" t will be to 1, and we are going g to have to make t "quite close" to 1 if we are going to guarantee that $\gamma(t)$ will be within only $1/10$ of $\gamma(1)$. A suitable choice is $\delta = 1/20$. It is indeed true that, whenever t is within $1/20$ of 1, $\gamma(t)$ is within $1/10$ of $\gamma(1)$. We thus conclude that it is true that for every positive ϵ, there exists a positive δ such that whenever $|t - 1| \leq \delta, d(\gamma(t), \gamma(1)) \leq \epsilon$ [namely, given positive ϵ, we choose $\delta = 1/2\epsilon$]. [We have not of course actually proven formally that this δ always works, but it is clear, and a full proof is not difficult.] We thus conclude that our curve is continuous at $t = 1$.

Is this curve continuous at $t = 5$? Yes, by exactly the same argument. Is it continuous at every t–value? Yes. So, this curve, since it is continuous at every $t = t_0$, is continuous.

Similarly, one verifies that the three curves illustrated on page 36 are all continuous. The curve γ with $\gamma(t) = (1, 1)$ if t is rational and $\gamma(t) = (4, 5)$ if t is irrational is not continuous at any $t = t_0$. [The curve "jumps between these two points every time t changes from a rational to an irrational and back". To show not continuous, one must find an ϵ such that the definition will not work. Since here the "skip", the distance between these points, is 5, one would choose for ϵ any number less than 5.]

6. Metric Spaces

So far, we have introduced four intuitive notions involving sets in the plane (or, if "continues curve" is included, five), expressed those intuitive notions as mathematical definitions, given various examples of those definitions, and proven a number of theorems which relate the definitions to each other. We are now at a crossroads: What shall we do next? One possibility would be to go ahead and look for more intuitive notions, more definitions and more theorems. It turns out, however that there is a more interesting and fruitful direction. It consists essentially of trying to isolate what is really relevant to what we already have, and trying to generalize it. It is the direction of "quality" as opposed to quantity. In this section, we begin this program.

Why have we been dealing, all this time, with the plane, as opposed to some other space? The reason, frankly, is that it is easy, for the plane, to draw figures to illustrated what is going on. Suppose, however, that one got interested in Euclidean three-dimensional space (the space of solid geometry)? A point of this space would be labeled by three numbers, so the coordinates of a typical point would be (x, y, z). One could, of course, introduce the notion of a set in this Euclidean three-dimensional space (an unambiguous statement of which (x, y, z) are in the set and which are out.). One would furthermore have similar intuitive ideas (of boundary, interior, bounded, and connected) about such sets. How would one go about capturing such intuitive ideas as definitions? The first step would be to observe that one also has a notion of geometrical distance in Euclidean three-dimensional space. Indeed, for $p = (x, y, z)$ and $q = (x', y', z')$ two points in this space, the distance between these points is given, by the Pythagorean theorem just as before, by

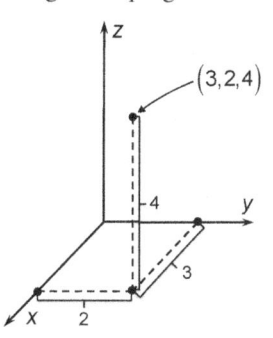

$$d(p, q) = \sqrt{(x' - x)^2 + (y' - y)^2 + (z' - z)^2}$$

So, one might decide, one will use this distance to formulate definitions of "boundary", "interior", "bounded", and "connected". One expects that finding suitable definitions this time will be somewhat easier, since we already have the experience gained with the plane.

But, were one to embark on such a program, a strange thing would happen. Not only does finding suitable new definitions turn out to be "somewhat easier" than it was before, it further turns out that one can just repeat, word for word, the old definitions, replacing "point of the plane" and "set in the plane" everywhere by "point of Euclidean three-dimensional space" and "set in Euclidean three-dimensional space". Thus, for example, the definition of "interior" on page 16 would become: "Let A be a set in Euclidean three-dimensional space. Then the *interior* of A is the set of all points p [now, of Euclidean three-dimensional space] such that, for some positive number ϵ, every point q [now, of Euclidean three-dimensional space] with $d(p,q) \leq epsilon$ is in A." The resulting four definitions not only make sense (i.e., are genuine mathematical definition), but furthermore accurately reflect the intuitive ideas they are to capture. One would thus instantly obtain all four definitions.

One would next seek theorems which relate these definitions. But now a second miracle takes place, One would soon realize that all our old theorems remain true for Euclidean three-dimensional space (with, again, the mere replacement of "plane" everywhere by "Euclidean three-dimensional space"). Not only this, but the old proofs are again proofs of the new theorems (with the usual replacement). Thus, for example, we would translate the theorem on page 26 as follows:

Theorem. Let A be a set in Euclidean three-dimensional space. Then every point of int (A) is in A.

Proof: Let point p in int (A). Then, for some positive number ϵ, every point within distance ϵ of p, is in A. But p is within distance ϵ of p and so p must be in A.

[Now, of course, "p" is a point of Euclidean three-dimensional space.]

In short, we have essentially already done this subject for Euclidean three-dimensional space.

So, one might turn to another possibility, say, one-dimensional space. The "coordinate" of a point would just be one number, (x), so our space is a single "line". Again, one can introduce sets in this one-dimensional space: again, one can introduce the geometrical distance between points (for $p = (x)$ and $q = (x')$ points in our one-dimensional space, set $d(p,q) = |x-x'|$, the geometrical distance between the points along the line); again, we can just repeat our old definitions (replacing "plane" by "line") to reflect our intuitive ideas of "boundary", etc.; again the old theorems remain theorems and the old proofs remain proofs. We have already done this subject for the line, too.

What would be one's response to this state of affairs? Purchase a large quantity of paper, and proceed to rewrite everything we have done two more times, but with this minor changes? Just say "similarly for Euclidean three-dimensional space, and for the line", we have above? The response of the mathematician would be the following. These "miracles" are an indication that we did not fully understand what we were doing in the first place. We were *talking* plane, but *doing* much more. A fuller understanding of what is really going on here is to come by somehow isolating what is genuinely relevant to our four notions and the relationships between them. What we want to do, then, is to generalize – to find a broader framework for our discussion which will at one stroke and encompass the plane, Euclidean three-dimensional space, the line, and hopefully even more possibilities.

What is the crucial thing which made our treatment of sets in the plane possible? What is the essential ingredient which our treatments of Euclidean three-dimensional space, the plane, and the line have in common? This is obvious: It is this distance d. Now comes the hard part. We have to figure out what it was about our distances that made things work out as they did for the plane: what the "distances" for Euclidean three-dimensional space, the plane, and the line have in common; what features our distances have which reflect our intuitive, geometrical ideas of what "distance" should be like. It is a problem not unlike that of inventing, for the first time, a definition of "boundary". It is, in the fullest sense, a problem in mathematics.

An example of an inappropriate property of distance is the following. For any two distinct points p and q, and any number c greater than $d(p, q)$, there are exactly two points r with $d(p, r) = c$ and $d(q, r) = c$. This property is true for the plane, Indeed, the two points lie on the perpendicular bisector of the line segment joining p and q, as shown in the figure. But this property does not hold is Euclidean three-dimensional space 9in which there is an infinite number of such "r' s "), or in the line (in which there is no such "r"). Furthermore, we never used this property, and it does not at all seem to be mandated by our intuitive idea of "distance". It is much more a property of "plane" than of "distance".

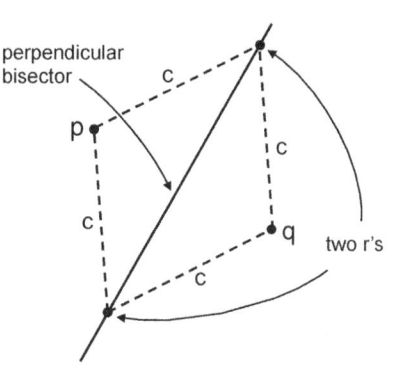

We now obtain the "distance-like" properties of our distances. The key to the first property is the theorem on page 26 (int $(A) \subset A$). What property of distance was used there? It was used in the part of the proof which reads "But p is within distance ϵ of p, ...". Of course, the reason why p is within distance ϵ of p is that $d(p, p) = 0$. Here, then, is a property of distance (that $d(p, p) = 0$) which is true for Euclidean three-dimensional space, etc. which was used in our treatment of the plane, and which seems natural for distance.

[If I have to accept "the distance of p from p is five units", then I do not want to think about that "distance".] A closely related property, which has been implicit in our discussion of the plane, is: If $p = q$ then $d(p,q)$ is positive. [That is, one cannot have distinct points with zero distance between them, or two points distance -7 apart.] The second property is such a simple and natural one that, while it has always been in the background, it has never been mentioned explicitly. [These are often the most difficult properties to find!] It is that, for any two points p and q, $d(p,q) = d(q,p)$. The distance from p to q is the same as the distance from q to p. Again, this property would seem to be an integral part of what we would wish to mean by "distance".

The third property was used, for example, in the theorem on page 28, in the sentence "Then every point within $1/2\epsilon$ of q must be within ϵ of p, ..." in the proof. Let r be such a point (within $1/2\epsilon$ of q). Then what is being used here is the following: If $d(p,q) \leq 1/2\epsilon$ and $d(q,r) \leq 1/2\epsilon$, then $d(p,r) \leq \epsilon$. Why is this true? Think of a triangle, with vertices p, q and r. Then if the distance from p to q is no more than $1/2\epsilon$ and the distance from q to r is no more than $1/2\epsilon$, the distance from p to r could hardly be more than $\epsilon(= 1/2\epsilon + 1/2\epsilon)$. It could hardly be that going from p to r via q is actually shorter than going "directly" from p to r. That is, it could hardly be that $d(p,q) + d(q,r)$ (distance from p to r via q) is less than $d(p,r)$ (distance from p to r directly). The property which is being used, then, is: $d(p,q) + d(q,r) \geq d(p,r)$. This is called the *triangle inequality*. It is true for the plane, for the line, and for Euclidean three-dimensional space. It was used in one of our theorems. Again, it seems like a natural property for "distance".

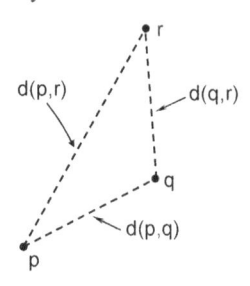

It turns out that an appropriate list of properties to reflect "distance" is just the three given above. In practice, of course, one would have to try many other properties, trying to make the decision, for each, whether or not it should be included. One would presumably also have second thoughts about these three. In any case, one eventually settles on these three properties. Having decided on the properties, one organizes everything as follows.

Definition. A *metric space* is a set X together with a rule which assigns, to any two points p and q of X, a number, $d(p,q)$, such that the following conditions are satisfied;

1. For any points p and q of X, $d(p,q)$ is zero if $p = q$ and positive if $p \neq q$.
2. For any points p and q of X, $d(p,q) = d(q,p)$.
3. For any points p, q and r of X, $d(p,q) + d(q,r) \geq d(p,r)$.

A metric space, then, is to represent "the idea of distance" in its purest form, with no other extraneous features". A metric space consists of a set

X (the "points between which distance will be given"), and the rule which assigns (unambiguously, of course) a number $d(p,q)$ to any two points p and q (which is to be regarded as, and will be called, the *distance* between p and q). But all these assignments of distances cannot be made just by whim, at least if one expects to obtain a metric space. They must be made in such a way that these three conditions are satisfied, such that they have at least some of the character of a "distance". Of course, to specify a metric space, one must say what the set X is, one must give some unambiguous rule for obtaining a number, $d(p,q)$, for *any* two points p and q of X, and one must have our three conditions satisfied. Nothing more and nothing less. In particular, it is not necessary that the metric space be "reasonable" in any other sense. [After all, we had the option, prior to the definition, to demand "reasonable", and in fact we took advantage of that option to impose our three conditions. Now, it is too late to ask for "still more reasonable". Of course, if one happened to come up with some nice additional conditions, nothing would stop one from defining a new thing to incorporate them.]

We now give some examples of metric spaces.

Let X be the plane, i.e., the set of all pairs, (x, y) of real numbers (certainly, a set). For p and q any two points of the plane, let $d(p,q)$ be the number given by the formula on page 10 (certainly, an unambiguous assignment of a number to any two points of X). Then our three conditions are of course satisfied, as we have already seen. This, then, is a metric space.

Let X be Euclidean three-dimensional space, i.e., the set of all triples, (x, y, z), of real numbers. For p and q any two points of X, let $d(p,q)$ be the number given by the formula on page 43. Our three conditions are certainly satisfied. This is a metric space. Similarly, the line (with $d(p,q)$ for $p = (x)$, $q = (x')$, given by $|x - x'|$) is a metric space.

Let X be the circle, i.e., the set of all pairs, (x, y), of numbers, with $x^2 + y^2 = 1$. Let, for p and q any two points of X, $d(p,q)$ be the geometrical distance between these points on the plane (i.e., use again the formula on page 10). Then the first condition is certainly satisfied ($d(p,q)$ is zero if $p = q$, and is positive otherwise); and the second condition id certainly satisfied (always, $d(p,q) = d(q,p)$). Finally, the third condition is also satisfied. Indeed, this is immediate, since points of X are "really" points of the plane (although, of course, not all points of the plane get to be in X), and the triangle inequality is satisfied for the plane.

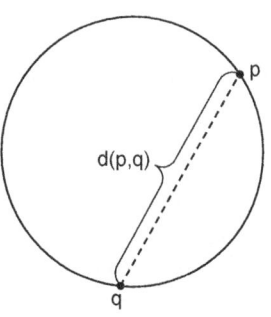

Let X be the same circle as on this page, and let, for p and q any two points of X, $d(p,q)$ be the length of the shortest segment of arc of the circle connecting p and q. Thus, for p and q as shown, $d(p,q)$ would be the length of the arc shown. [Note that this is an unambiguous assignment which yields a number in exchange for two points.] Again, the first two conditions are immediate. One furthermore easily convinces oneself that the third condition is also satisfied. We have a metric space. [Which one is right, this metric space or the previous one? What is the real distance between points on the circle? There is no "right" or "real" in mathematics. To give a metric space, one must give a set X and the distances. That is it. The definition does not require that anything be "real".]

Let X be the plane. For p and q any two points of X, let $d(p,q)$ be the number which is zero if $p = q$, and one if $p \neq q$. [This is an unambiguous, if "unreasonable", rule. Thus, $(0,0)$ and $(.0001, 0)$ are distance 1 apart: $(0,0)$ and $(23,926, -189)$ are distance 1 apart.] Let's see if the conditions are satisfied. For $p = q$, $d(p,q)$ is certainly zero. For $p \neq q$, $d(p,q)$ is one which is certainly positive. The first condition is ok. furthermore, $d(p,q) = d(q,p)$ (since both sides are zero if $p = q$ and both sides are one if $p \neq q$). Finally, we come to the third condition. How could this possibly be violated? Since all "distances" are either one or zero, the only possibility would be if $d(p,q)$ is zero, $d(q,r)$ is zero, while $d(p,r)$ is one. But this could never happen, for $d(p,q) = 0$ would mean $p = q$, and $d(q,r) = 0$ would mean $q = r$ – and so we would have to have $p = r$, and so $d(p,r) = 0$ also. We conclude, then, that even the triangle inequality is satisfied. This is a metric space!

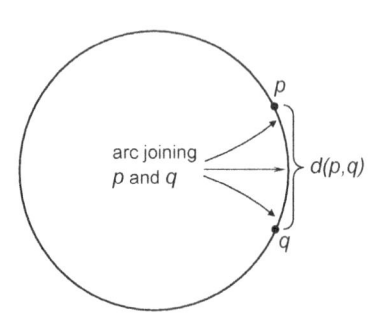

Let X be the plane. For p and q any two points of X, let $d(p,q)$ be the usual geometrical distance between p and q if that distance is less than one, and one if that geometrical distance is greater than or equal to one. The distances (with this d) of some points from p are shown in the figure. Arguing as in the previous example, one sees that this is a metric space. [Again, the only tricky step is the triangle inequality. One must consider several cases, according to whether the geometrical distances between p, q and r are less than or greater than one.]

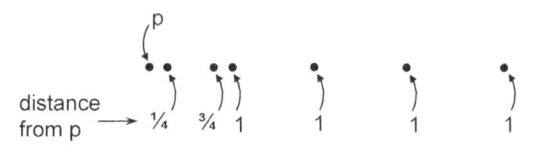

Let X be the plane. Let, for p and q any two points of X, $d(p,q)$ be five times the geometrical distance between p and q. This is a metric space.

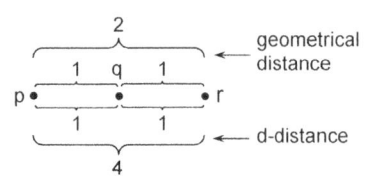

Let X be the plane. Let for p and q

any two points of X, $d(p,q)$ be the square of the geometrical distance from p to q. This "distance" does satisfy the first two conditions for a metric space. However, it does not satisfy the third condition. Let, for example, p, q and r be as shown (so the geometrical distance from p to q, and from q to r, are both one). Then we would have $d(p,q) = 1^2 = 1$, $d(q,r) = 1^2 = 1$, but $d(p,r) = 2^2 = 4$. It is not true that $d(p,q) + d(q,r) \geq d(p,r)$. This is not a metric space.

Let X be the plane. Place a unit sphere on X, as shown in the figure, and let N denote the "North pole". For p and q any two points of the plane, we obtain a number $d(p,q)$ as follows. First draw a straight line from p to N, and identify the point p' at which that line enters the sphere. Similarly, identity the point q'. Now take the straight-line distance (in our three-dimensional space of the figure) between p' and q'. This distance is to be the number $d(p,q)$. Note that this is indeed a rule to get a number from any two points of X. Points "very far out near the edge of the plane", e.g., (23, 234, 9.001), yield points p' "very near the North pole N". Thus, any two such points will, from our d, be "very close together", even though they appear geometrically to be very far apart. That the first two conditions are satisfied is immediate. It actually turns out (although it is perhaps a bit complicated to check in detail) that the triangle inequality is satisfied. This is a metric space.

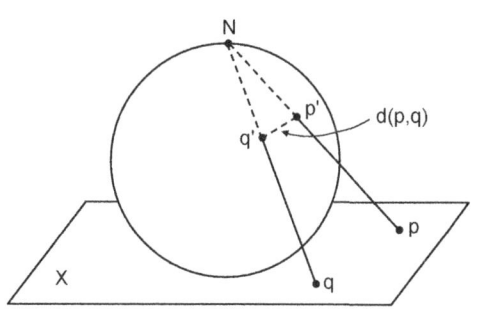

Let X be the set consisting of the one point p, and set $d(p,p) = 0$. This is a set and a rule. The three conditions are easily checked. This is a metric space. Let X be the set consisting of just two points, p and q. Set $d(p,p) = 0$, $d(q,q) = 0$, $d(p,q) = 19$. This is a metric space.

Let X consist of just four points, with distances as indicated in the figure, and with $d(p,p)$ always zero. There are a total of sixty-four possibilities for p, q and r in the triangle inequality. One checks them. [It is actually easy: Almost all the cases reduce to just a couple.] The first two conditions are immediate. This is a metric space. If one of the "17"s above were changed to a "15", one would not have a metric space, for the triangle inequality would be violated. [For which p, q, r?]. If one of the "17"s were changed to a "0", the first condition would also be violated, and again we would not have a metric space.

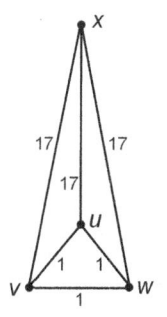

We now consider a couple of more exotic examples.

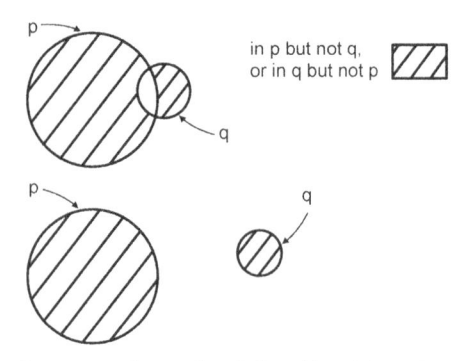

Let X be the set of all disks in the plane. So, a "point" of X is actually an entire disk. Let p and q be two "points" of X as shown. We obtain $d(p,q)$ as follows. Consider the set of all points of the plane which are in p (a disk) but not in q, or else are in q but not in p, as in the examples shown above. Then, let $d(p,q)$ be the area, in the plane, of this set of points. Thus, in the first example on the right, $d(p,q)$ is rather large, for the area of the region "in p but not q or in q but not p" is rather large. In the second example, $d(p,q)$ is rather small, i.e., these "points" of the set X are "close together". This X, d we claim, is a metric space. For the first condition, note that $d(p,q)$, as the area of some region in the plane, is always greater than or equal to zero.

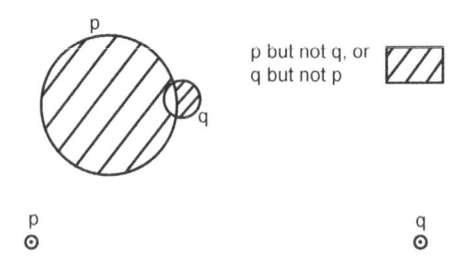

Furthermore, in order for $d(p,q) = 0$ there would have to be no points in p but not q, and no points in q but not p, i.e., p and q (points of X) would have to be the same disk in the plane. That $d(p,q) = d(q,p)$, the second condition, is immediate.

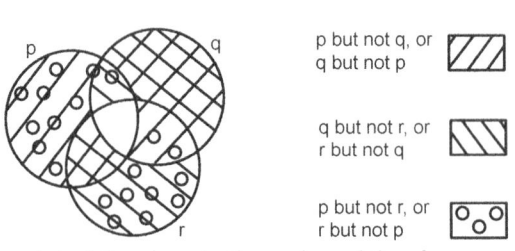

Finally, we check the triangle inequality, $d(p,q)+d(q,r) \geq d(p,r)$. How each of the "$d$'s" above of course represents the area of some region in the plane. In order to show this inequality, therefore, it suffices to show that every point of the plane in the region of the plane corresponding to $d(p,r)$ is either in the region corresponding to $d(p,q)$ or in the region corresponding to $d(q,r)$ (or, of course, possibly in both). Having shown this, it will follow that the area one takes to compute $d(p,r)$ cannot exceed the sum of the areas for $d(p,q)$ and $d(q,r)$. Let, then, u be a point in $d(p,r)$, so that u is either in p but not r or in r but not p. We must show that this u is either in the region for $d(p,q)$ or in the region for $d(q,r)$. There are two cases to consider: u in p but not r and u in r but not p. Let us take the first, so u is in p but not r. We now ask: Is u in the set q? If yes, then u (in p

but not r, and assumed to be in q) is in q but not r – and so is in the region whose area represents $d(q, r)$. If no, then u (in p but not r, and assumed not to be in q) is in p but not q – and so is in the region whose area represents $d(p, q)$. Similarly for the case u in r but not p. We conclude, then, that this is a metric space. [In terms of the figure above, what we have shown is that "the entire region with circles in it has stripes going one way or the other".]

Similarly, one could let X be the set of all rectangles in the plane, and define d as above, using the area. Or, one could let X consist of disks and rectangles, and again define d using the area. In either case, we obtain a metric space. [Why not just let X be the set of all subsets of the plane, and again define d by using the areas? This will not work, for it would be necessary to be able to say what the "area" of an arbitrary subset of the plane is. What, for example, is the "area" of the set consisting of all (x, y) with x rational?]

Let X be the set of all finite subsets of the positive integers, so a "point" of X is just a finite list of integers, with no repetitions. For example, $p = (19, 3, 197, 11)$ is a point of X. Now consider two such points of X, say this one and $q = (197, 11, 5)$. We compute $d(p, q)$ as follows. First, strike out any integers which appear in both p and q (in this example, 197 and 11). Then add all the remaining integers (in this example, 19, 3, and 5). The resulting sum is $d(p, q)$, so in this example $d(p, q) = 19 + 3 + 5 = 27$. One verifies that this is a metric space. [The argument is essentially the same as for the example above.]

Let X be the set of all ordered pairs of points in the plane, so a point of X, $p = (u, v)$, consists of a first point, u, of the plane and also a second point, v of the plane. [We permit $u = v$.] Given two such pairs, say the one above and $q = (u', v')$, we define the distance between these two points of X by $d(p, q) =$ (geometrical distance between u and u') + (geometrical distance between v and v'). Thus, in order that two points of X be "close together", the corresponding first points of the plane must be close together in the plane, and also the corresponding second points. One easily checks, using for each of the three conditions the corresponding condition for the plane, that this X, d is a metric space.

7. Boundary, Interior, Bounded, and Connected in Metric Space

We decided that, in our discussion of properties of sets in the plane, we were talking plane, but actually doing much more. Our attempt to better understand what was going on, to focus on what was really relevant to our discussion, led to the notion of a metric space, "distilled essence of distance". We have some examples of metric spaces – enough examples to see that, while "metric space" does in some sense capture our ideas of distance, it also admits some pretty strange-looking things. The next step is to go ahead and carry out the program for which our metric spaces were invented. We want to repeat our treatment of sets in the plane, but now for "sets in a metric space". We want to carry out this treatment once and for all – for a general metric space. Then the plane, Euclidean three-dimensional space, etc. will all be special cases. In this section, we carry out the first half of this program: We introduce the definitions, with examples, of "boundary", "interior", "bounded", and "connected" in a metric space. One expects, of course, that the new definitions will be essentially the same as the old ones, since it was precisely the fact that our earlier discussion seemed to have a wider domain of applicability which led us in the direction of a metric space in the first place.

We begin with "boundary" and "interior". We have

Definition. Let X, d be a metric space, and let $A \subset X$. Then the *boundary* of A is the set of all points p of X such that, for every positive number ϵ, there is a point q of X with $d(p,q) \leq \epsilon$ and q in A, and also a point q' of X with $d(p,q') \leq \epsilon$ and q' not in A.

Definition. Let X, d be a metric space, and let $A \subset X$. Then the *interior* of A is the set of all points p of X such that, for some positive number ϵ, every point q of X with $d(p,q) \leq \epsilon$ is in A.

Note that these are just our earlier definitions, but with a few minor changes. Instead of "A is a set in the plane", we say "A is a subset of X", or $A \subset X$". [In the case in which X *is* the plane, a "set in the plane" is the same thing as a "subset of X".] Further, we include for emphasis the fact that all points considered in the definition are supposed to be points of X. We shall continue to write bnd (A) and int (A) for the boundary and interior of A.

We give some examples.

Let X be the plane, and d the usual geometrical distance. Then these definitions of course reduce to our old definitions of the boundary and interior. Thus, for any set A in this metric space, bnd (A) and int (A) are the boundary and interior we have found before.

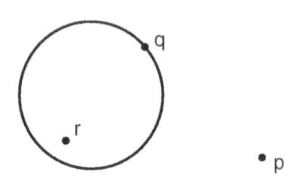

Let X be the plane, and let d be: $d(p, q) = 1$ if $p \neq q$, and $d(p, q) = 0$ if $p = q$. This, as we have seen, is a metric space. Let A be the usual disk (all (x, y) with $x^2 + y^2 < 1$). We determine int (A). Consider first the point p (given, say, by (3,4)). Is p in int (A)? Is it true that, for some positive ϵ every point of X within distance ϵ of p is in A? It is not true. Indeed, no matter what positive ϵ is chosen the point p itself will be within ϵ of p, while of course p is not in A. So, this p is not in int (A). Similarly, the point q shown in the figure is not in int (A). What about the point r? Is it true that, for some positive ϵ, every point of X within ϵ or r is in A? It is true. Indeed, consider $\epsilon = 1/2$ (certainly, *some* positive ϵ). What is the locus of points within 1/2 of r? Well, the distance of a point from r is 1 if that point is different from r, and 0 if that point is r itself. Thus, the locus of points within 1/2 of r is just r itself. [Note; It is *not* a disk of radius 1/2 centered at r. For "distance", we must of course use the "*d*" of our metric space.] So, every point within 1/2 of r (since there is only one, namely r) is in A. So, r is in int (A). Clearly, then, int (A) = A, i.e., we get the same answer as we obtained before. What is bnd (A)? Is point p in bnd (A)? Is it true that, for every positive ϵ, there is a point within ϵ of p and in A, and also a point within ϵ of p and not in A? It is not true. Consider, for example, $\epsilon = 1/2$. The locus of points within 1/2 of p consists of just p itself. Since p is not in A, there is no point within 1/2 of p and in A. So, p is not in bnd (A). Is r in bnd (A)? Is it true that, for every positive ϵ, there is a point within ϵ of r and in A, and also a point within ϵ of r and not in A? Again, it is not true: Chose $\epsilon = 1/2$, and so the locus of points within 1/2 of r consists of r itself, and so there is no point within 1/2 of r and not in A. Finally, we consider the point q. It is true that, for every positive ϵ, there is a point within ϵ of q and in A, and also a point within ϵ of q and not in A? This is not true either! Try $\epsilon = 1/2$. The locus of points within 1/2 of q consists of q itself. But q is not in A. So, there is no point within 1/2 of q and in A. Clearly, then, bnd (A) is the empty set.

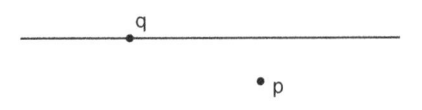

Next, let, in this metric space, A be the line, the set of all (x, y) with $y = 1$. Is point p in int (A). Can we find a positive ϵ such that every point within ϵ of p is in A? Certainly not: point p will, no matter what positive ϵ is chosen, be within ϵ of p, while p is not in A. Is point q in int (A)? It is. Choose $\epsilon = 1/2$. Then the locus of points within $1/2$ of q is just q itself. Thus, *every* point $1/2$ of q is in A. Thus, this q is in int (A)! Clearly, int $(A) = A$, i.e., is just the line. What about bnd (A)? Is point p in bnd (A) It is not: For $\epsilon = 1/2$, there is no point within ϵ of p and in A. Similarly, point q is not in bnd (A): For $\epsilon = 1/2$, there is no point within ϵ of q and not in A (since, of course, the locus of points within $1/2$ of q consists of just q itself]. Thus, bnd (A) is the empty set. [In these examples, we used "1/2" because it is positive and less than 1. For p a point of X, the locus of points within 2 of p is of course the entire set X.]

We notice a pattern in the examples above. In each case, we had int $(A) = A$ and bnd (A) the empty set. Inspection of the arguments also reveals that the arguments for the two choices of A were essentially the same. In fact, these arguments had virtually nothing to do with A. These observations suggest

Theorem. Let X, d be the metric space with X the plane and $d(p, q) = 1$ if $p \neq q$ and $d(p, q) = 0$ if $p = q$. Thus, for any $A \subset X$, int $(A) = A$ and bnd (A) is the empty set.

Proof: For p in A, every point within $1/2$ of p is in A, and so p is int (A). For p not in A, there could be no positive ϵ such that every point within ϵ of p is in A, for p itself, not in A, will always be within ϵ of p, and so p is not in int (A). So, int $(A) = A$. For p any point of X, the set of points within $1/2$ of p consists of p itself. So, it cannot be the case that for every positive there is a point within ϵ of p and in A, and also a point within ϵ of p and not in A. So, p is not in bnd (A). Hence, bnd (A) id the empty set.

What is going on here? Why do we get answers so different from what we might have expected? Are we to conclude that there is something wrong with the definitions? Our conclusion should be something quite different. What the theorem is trying to tell us is that our intuitive ideas about the plane are intuitively connected with geometrical distance in the plane. Whether we realized it or not, geometrical distance was an integral part of our intuition about "interior" and "boundary". Change the distance (e.g., to obtain the metric space above), and the answers change. It is through the introduction of metric spaces, then, that we come to see the constituents of our ideas about sets in the plane.

Next, let X be the plane, and let d be: $d(p, q)$ is the geometrical distance from p to q if this geometrical distance is less than one, and $d(p, q) = 1$ if that geometrical distance is greater than or equal to 1. This is of course a metric space. The locus of points within distance $1/2$ of point p consists of a disk of radius $1/2$ centered at p; the locus of points within distance $3/4$ of p is a disk of radius $3/4$ centered at p. But, the locus of points within distance 2 of p is the entire plane (since all "distances" in this metric space are less than or equal to 1). Now let A again be our disk. [See page 54] Is point p in int (A)? It is not. No matter what positive ϵ is chosen, point p will be within distance ϵ or p, while p is not in A. Is point q in int (A)? Again no, for the same reason. What about point? Is there a positive ϵ such that every point within ϵ of r is in A? There certainly is, say, $\epsilon = 1/10$. The locus of points within $1/10$ of r is the disk of radius $1/10$ centered at r (since $1/10$ is less than one), while all points in this disk are certainly in A. So, r is in int (A). Clearly, int $(A) = A$, i.e., the same answer we got before. What about bnd (A)? Point p is not in bnd (A), for say, for $\epsilon = 1/3$ (so the locus of points within $1/3$ of p is the disk of that radius centered at p), there is no point within $1/3$ of p in A. Point r is not in bnd (A) (for there is no point within $1/10$ of r and not in A). What about point q? Is it true that, for every positive ϵ, there is a point within ϵ of q and in A, and also a point within ϵ of q and not in A? Let us try some ϵ's to see if they work. What about $\epsilon = 5$. Is there a point within 5 of q in A, and also a point within 5 of q and not in A? The locus of points within 5 of q is the entire plane, and so there certainly is. Clearly, there also is for $\epsilon = 10$, or $\epsilon = 2$, or, indeed, for any ϵ greater than 1. What about something smaller than 1, say, $\epsilon = 1/2$. Is there a point within $1/2$ of q and in A, and also a point within $1/2$ of q and not in A? The locus of points within $1/2$ of q is the disk of that radius centered at q. Clearly, there are such points. One now sees that this will work for *every* positiveϵ. Hence, point q is in bnd (A). Clearly, bnd (A) is the "rim" of the disk, i.e. the set of all points (x, y) with $x^2 + y^2 = 1$.

Next, let A be the line as illustrated at the end of page 54 As usual, the point p is not in int (A). Is q in int (A)? We must try to find a positive ϵ such that every point within ϵ of q is in A. Will $\epsilon = 5$ do? No, for the locus of points within 5 of q is the entire plane, and not all these points are in A. Will $\epsilon = 1/2$ do? No, for the locus of points within $1/2$ of q is the disk of that radius centered at q, and not all these points are in A. Clearly, we are not going to find such an ϵ. So, point q is not in int (A) either. Thus, int (A) is the empty set. Is point p in bnd (A)? It is not. For, say, $\epsilon = 1/10$, so the locus of points within $1/10$ of p is the disk of that radius, there will be no point within ϵ of p and in A. Is point q in bnd (A)? Is it true that, for every positive ϵ, there is a point within ϵ of q and in A, and also a point within ϵ of q and not in A? We try a few ϵ's. Consider $\epsilon = 5$, so the locus of points within 5 of q is the entire plane. There certainly is a point within 5 of q and in A, and also a point within 5 of q and not in A. Clearly, $\epsilon = 10$, or $\epsilon = 2$ will work, too. What about, say, $\epsilon = 1/2$? Is there a point within $1/2$ of q and in A, and also a point within $1/2$ of q and not in A? The locus in this case is a disk of radius $1/2$, and certainly there are such points. Clearly, this will work for *any* positive ϵ. So, point q is in bnd (A). We conclude, then, that bnd $(A) = A$, i.e., the same answer as we obtained before for the usual plane.

Again, we see a pattern to these examples. In both cases, we got the same answer as we obtained earlier for the ordinary plane. Looking back at the arguments, one realizes that this is no coincidence. The mechanism is the following. Let us written, as above, d for the distance in the present metric space, and d' for ordinary geometrical distance in the plane. Then, for any points p and q of X, $d(p,q)$ is just $d'(p,q)$ if the latter is less than or equal to one, and is one if $d'(p,q)$ is greater than one. Now fix a subset A of X. Let us write "int" for the interior using distance d, and "int'" for the interior using geometrical distance d'. We want to show that int $(A) = int'(A)$. Consider, then, some point p of int (A). This means of course that there is some positive ϵ such that every q with with $d(p,q) \leq \epsilon$ is in A. We want to show that this p is also in int' (A). That is we want to find some positive number ϵ' such that every q with $d'(p,q) \leq \epsilon'$ is in A. So, we have ϵ (such that every q with $d(p,q) \leq \epsilon$ is in A), and we are looking for ϵ' (such that every q with $d'(p,q) \leq \epsilon'$ is in A). The question is: What should we choose for ϵ'? To see the answer, recall that $d(p,q)$ is just $d'(p,q)$ if the latter is less than or equal to one, and one otherwise. Clearly, then, we always have $d(p,q) \leq d'(p,q)$, i.e., the effect of our "funny" distance is to regard points as at least as close together as they are geometrically. Because of this, we claim we can just choose $\epsilon' = \epsilon$ and it will work. Does this choice of ϵ' satisfy our condition? Is it true that, every q with $d'(p,q) \leq \epsilon'$ is in A? Yes, for if $d'(p,q) \leq \epsilon'$, then (since $d(p,q) \leq d'(p,q)$), we also have $d(p,q) \leq \epsilon'$, and so (since we choose $\epsilon' = \epsilon$) we also have $d(p,q) \leq \epsilon$, and so (by the defining property of ϵ) we then have q in A. Thus, we supposed that p was in int (A), i.e., we supposed a positive ϵ such that every q with $d(p,q) \leq \epsilon$ is in A. We then found an ϵ' (namely, ϵ) such that every q with $d'(p,q) \leq \epsilon'$ is in A. That is, we showed that p is also in int' (A). That is, we have shown that every point of int (A) is also in int' (A).

We now do the reverse. We fix a point p in int' (A), and we want to show that this p must be in int (A). That is, we have positive number ϵ' (such that every q with $d'(p,q) \leq \epsilon$ is in A), and we are looking for ϵ (such that every q with $d(p,q) \leq \epsilon$ is in A. What should we choose for this ϵ? What about choosing $\epsilon = \epsilon'$? Let us try an example, say $\epsilon' = 5$. For this example (choosing $\epsilon = \epsilon'$) we would have that every q with $d'(p,q) \leq 5$ is in A, and we wish to conclude that every q with $d'(p,q) \leq 5$ is in A. Does this follow? Well, the locus of points with $d'(p,q) \leq 5$ is a disk of radius 5 centered at p (for d' is just geometrical distance). However, the locus of points q with $d(p,q) \leq 5$ is the entire plane (for, no matter what q is, $d(p,q)$ can be at most one, and so will always be less than or equal to 5). Thus, we would have that every point of the disk of radius 5 about p is in A, and would wish to conclude that every point of the plane is in A. Clearly, this does not follow. Thus, we have ϵ' (such that every q with $d'(p.q) \leq \epsilon'$ is in A, and are looking for ϵ (such that every q with $d(p,q) \leq \epsilon$ is in A. We tried making the choice

$\epsilon = \epsilon'$, and found, at least for $\epsilon' = 5$, that it will not work. What ϵ *will* work for $\epsilon' = 5$? We claim that $\epsilon = 1/2$ will work. The locus of points with $d(p, q) \le 1/2$ is the disk of that radius centered at p, and all these points are within the disk of radius 5 about p. Thus, given that every q with $d'(p, q) \le 5$ is in A, it follows that every q with $d(p, q) \le 1/2$ is in A. So, for this ϵ' (namely 5), we have found a suitable ϵ (namely 1/2). Let us try another ϵ', say $\epsilon' = 137$. For this choice we would have that every q with $d('p, q) \le 137$ is in A, and we are looking for ϵ such that every q with $d(p, q) \le \epsilon$ is in A. Again, $\epsilon = 1/2$ will do, for every q with $d(p, q) \le 1/2$ (i.e., every q in the disk of radius 1/2 centered at p) will have $d'(p, q) \le 137$ (i.e., will be in the disk of radius 137 centered at p), and so will be in A. Clearly, then, $\epsilon = 1/2$ will do for these "large" ϵ'. Will $\epsilon = 1/2$ work for *any* positive ϵ'? Let us try $\epsilon' = 1/10$. For this choice, we would have that every q with $d'(p, q) \le 1/10$ is in A, and we wish to conclude that every q with $d(p, q) \le 1/2$ is in A. Is this true? The locus of points q with $d'(p, q) \le 1/10$ is the disk of that radius centered at p: the locus of points q with $d'(p, q) \le 1/10$ is the disk of that radius centered at p. Can we conclude, knowing that every q in the disk of radius 1/10 centered at p is in A, that every q in the disk of radius 1/2 centered at p is in A? We cannot. So, this ϵ (namely, 1/2) will not work for $\epsilon' = 1/10$. But clearly, an ϵ which *will* work for $\epsilon' = 1/10$ is $\epsilon = 1/10$.

The suitable is now clear. We have ϵ' (such that every q with $d'(p, q) \le \epsilon'$ is in A), and are looking for ϵ (such that every q with $d(p, q) \le \epsilon$ is in A). When ϵ' is "large" (specifically, greater than one), we must choose ϵ something like 1/2 (for if we dare choose ϵ also large, specifically, greater than one, then the locus of points q with $d(p, q) \le \epsilon$ will become the entire plane, and we shall not be able to conclude that these are all in A). However, when ϵ' is "small" (specifically, less than one), then we must start making ϵ small too (for if we were to continue choosing $\epsilon = 1/2$, then the locus of points q with $d(p, q) \le \epsilon$ will continue to be the disk of radius 1/2 centered at p, and this disk will not remain within the locus of points q with $q'(p, q) \le \epsilon'$). A suitable choice for our ϵ, then, is this: For $\epsilon' \ge 1$, choose $\epsilon = 1/2$; for $\epsilon' < 1$, choose $\epsilon = \epsilon'$. For this choice, it will indeed be the case that, given that every q with $d'(p, q) \le \epsilon'$ is in A, every q with $d(p, q) \le \epsilon$ is in A. Thus, we have shown that every point of int$'(A)$ is also in int(A).

Putting our two conclusions together, we conclude that int $(A) = \int'(A)$, for any set A in the plane. The key observations which led us to this conclusion were i) given positive ϵ such that every q with $d(p, q) \le \epsilon$ is in A, there exists positive ϵ' (namely, $\epsilon' = \epsilon$) such that every q with $d'(p, q) \le \epsilon'$ is in A, and ii) given positive ϵ' such that every q with $d'(p, q) \le \epsilon'$ is in A, there exists positive ϵ (namely, $\epsilon = 1/2$ if $\epsilon' \ge 1$ and $\epsilon = \epsilon'$ if $\epsilon' < 1$) such that every q with $d(p, q) \le \epsilon'$ is in A. By a similar argument, one concludes that bnd (A) (the boundary using distance d) is always the same as bnd$'(A)$ (the boundary using distance d').

What is one to make of all this? We began with the boundaries and interiors of sets in the plane, using the usual geometrical distance. We now try "changing the distance". For one such change (making the distance zero if the points are the same, and one otherwise), our definitions yield strange answers, which do not at all correspond with our intuition. But for another change (making the distance the geometrical distance if that is less than or equal to one, and one otherwise), our definitions yield precisely the boundary and interior we obtained before. What these examples are doing is giving us detailed information about what facets of "distance" are actually relevant to our intuition of "boundary" and "interior". It is clear that not *everything* about distance counts for the boundary and interior, for at least one significant change in what these distances are to be changes neither the boundary nor the interior. On the other hand, *something* about distance counts, for another change in what the distances are to be changes both the boundary and interior. What, exactly, is it about "distance" which *is* relevant to our ideas of boundary and interior? This is a question to which we shall return shortly.

We give a few more examples of interiors and boundaries of sets in metric spaces.

Let X, d be the metric space of page 50, so X is the set of all disks in the plane, and d is computed using areas. Let A be the subset of X consisting of all disks which lie entirely within some fixed square in the plane (say, the square given by all (x, y) with $0 \le x \le 1$ and $0 \le y \le 1$). Thus, for example, the disks p and s in the figure are not in A. While disk r is in A. [Note that A is indeed a subset of X. We have clearly specified which points of X are in A and which are not.] What is int (A)? The point p of X will certainly not be in int (A): It is not even close to being true that, for some positive ϵ, every point q (disk!) within distance (using areas!) ϵ of p is in A. The point s of X is also not in A. Given positive ϵ, let q be a disk obtained by displacing s very slightly to the left. Then, if this "very slightly" is small enough, q will be within distance ϵ of s. But, of course, this q (no matter how small "very slightly" is) will not be in A. So, it is not true that, for some positive ϵ, every q with $d(q, s) \le \epsilon$ is in A. So, s is not in int (A). The point r is in int (A). Clearly, then, int (A) will consist of all disks which lie within the square, and whose edge does not touch the boundary of the square. On the other hand, bnd (A) consists of all disks which lie within the square, and whose edge does touch the boundary of the square (for, in order, that disk p be in bnd (A), it must be the case that for every positive ϵ there is a disk within ϵ of p and in A, and also a disk within ϵ of p and not in A).

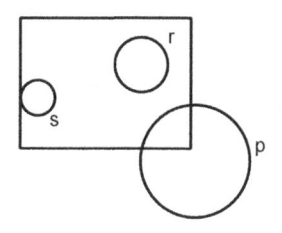

Let the metric space be the same, but now let A be the subset of X consisting of all disks with center at the origin, $(0, 0)$. What is int (A)? Clearly, the only serious candidates for points in int (A) are disks with center the origin. Is such a disk in int (A)? It is not. No matter what positive ϵ is given, let q be obtained from p (the one with center the origin) by a small displacement, so q will be within ϵ of p, but q will not be in A. So, int (A) is the empty set. On the other hand, bnd (A) = A.

Let X be the set consisting of all points (x, y) of the plane with x rational. Let d be the usual geometrical distance. This is a metric space. Let A be the subset of X consisting of all points *of x* with $x > \pi$. Thus, the point (20, 17) is in A, while $(-1, 17)$ is not (because we do not have $x > \pi$), and $(2\pi, 17)$ is not (because this "point" is not even in X: It does not have x rational. What is int (A)? The point p, given by $(-1, 1)$ is not even close to being in int (A). What about the "point" r, given by $(\pi, 1)$? This is not in int (A), because it is not even in X. It is not even a "point" of the metric space we are considering. Finally, consider the point s, given by (4, 1). Is it true that, for some positive ϵ, every point q of X with $d(p, q) \leq \epsilon$ is in A? This is true! Choose $\epsilon = 1/2$. Then every point q of x with $d(p, q) \leq 1/2$ is indeed in A. [What about the point $q = (1.2\pi, 1)$ which would seem to be within 1/2 of s, and not in A? But this is not a "point" of our metric space X, because its x-value is not rational.] Clearly, then, int (A) = A. What is bnd (A)? Point p is certainly not in bnd (A). "Point" r is not even under consideration. What about point s? This point s is not in bnd (A), for it is not true that, for every positive ϵ, there is a point within ϵ of s and not in A. Indeed, choose $\epsilon = 1/2$. There is, as we just saw, no point q of X within 1/2 of s and not in A. So, in this example, bnd (A) is the empty set.

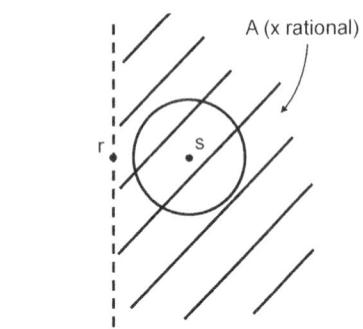

Let the metric space be the same, and let A be the line given by all (x, y) with $x = 1$. Then int (A) is the empty set, and bnd (A) = A. Let A be "the closest thing we can have to our disk": All (x, y) with $x^2 + y^2 < 1$ and x rational (a subset of X). Then int (A) = A, while bnd (A) consists of all (x, y) with $x^2 + y^2 = 1$ and x rational.

These, then, are some examples of the boundaries and interiors of sets in metric space.

We turn next to the definition of "connected" for a set in a metric space. Again, the appropriate definition is suggested by the corresponding definition (page 20) for a set in the plane.

<u>Definition</u>. Let X, d be a metric space, and let $A \subset X$. Then A is said to be *disconnected* if there exists a subset B of X such that some point of A is in B, some point of A is not in B, and no point of A is in bnd (B).

Note that we deal only with the possible disconnectedness of a subset of X. We also require that B be a subset of X. Of course, "bnd (B)" above is the boundary using the metric space X, d. We give some examples.

Let X be the plane, and d ordinary geometrical distance. Then this definition of course reduced to our earlier definition (of disconnected for a set in the plane). Thus, our earlier examples of connected and disconnected sets in the plane apply.

Let X be the plane, and let d be: $d(p, q) = 1$ if $p \neq q$ and $d(p, q) = 0$ if $p = q$ – a metric space. Let, for example, A be the disk: the set of all (x, y) with $x^2 + y^2 < 1$. Is A disconnected? Yes. Let, for example, B be the set of all points (x, y) with $x > 0$, as illustrated on the right. Then some point of A (for example, $(1/2, 0)$) is in A, and some point of A (for example, $(-1/2, 0)$) is not in A. Furthermore, no point of A is in bnd (B), for as we have seen (theorem on page 55) bnd (B) must be the empty set, and so no point whatever is in bnd (B). Thus, A is disconnected. Clearly, the key feature which causes this result is the fact that the boundary of any subset of this metric space is the empty set. In fact, something much more general is true. Let A be any subset of X (in this metric space) having at least two points, say p and q. Then, we claim A must be disconnected. [Choose for B the subset consisting of just the point q. Then some point of A is in B (namely, q), some point of A is not in B (namely, p), and no point of A is in bnd (B) (since bnd (B) is the empty set). Thus, any subset of X having two or more points must be disconnected. On the other hand, a subset of X consisting of just one point, as will as the empty subset of X, are connected.

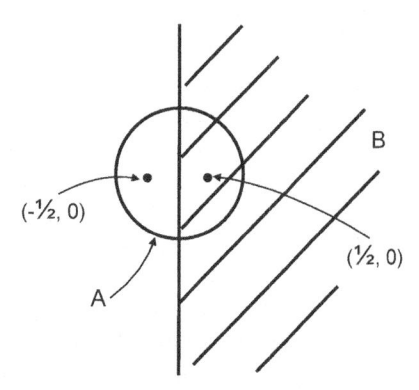

Nest, let X be the plane, and let $d(p, q) =$ the geometrical distance from p to q if that distance is less than one, and $d(p, q) = 1$ if that geometrical distance is greater than or equal to one. Which subset of X will be disconnected in this metric space? The answer, as it turns out, is extremely simple. First recall that, for any subset B of X, bnd (B) (for this metric space) is precisely the same set as boundary one would get if the distance were instead the ordinary geometrical distance (what we shall call bnd' (B)). It follows from this, we claim, that the disconnected sets in this metric space are exactly the disconnected sets in the ordinary plane. Indeed, suppose that $A \subset X$ is disconnected in this metric space, so there is some $B \subset X$ such that some point of A is in B, some point of A is not in B, and no point of A is in bnd (B). Then (since bnd $(B) =$ bnd'(B)), it follows that some point of A is in B, some point of A is not in B, and no point of A is in bnd' (B). That is, it follows that A is disconnected in the plane with the usual geometrical distance. For the reverse, suppose that $A \subset X$ is disconnected with the usual geometrical distance, so there is a set $B \subset X$ such that some point of A is in B, some point of A is not in B, and no point of A is in bnd' (B). Then it follows that some point of A is in B, some point of A is not in B, and no point of A is in bnd (B). That is, it follows that A is disconnected also in this metric space. In short, since this "chop the distances' down to one" metric space gives the same boundaries as the ordinary plane, it also gives the same connected and disconnected sets.

Let X be the set consisting of all points (x, y) of the plane with x rational. Let d be the usual geometrical distance. Set $A = X$ (certainly a subset of X). Is A disconnected? The answer is yes. Let $B \subset X$ consist of all points of X with $x > \pi$. Then certainly some point of A is in B (since $A = X$), some point of A is not in B, and yet (since bnd (B) is the empty set) no point of A is in bnd (B). Next, let A be "the closest thing to a disk" in this metric space: all (x, y) with $x^2 + y^2 < 0$ and x rational. Is A disconnected? It is. Let $B \subset X$ consist of all points (x, y) of X with $x > \pi/4$. [Where did we get this number "$\pi/4$"? It has to be an irrational number (in order to ensure that bnd (B) will be the empty set. For example, if B were all (x, y) in X with $x > 1$, then bnd (B) would be al (x, y) in X with $x = 1$ – not the empty set). Furthermore, it has to be less than one (to ensure that some point of A is in B), and also greater than minus one to ensure that some point of A is not in B). So, any irrational between -1 and 1 would do.] Finally, let A consist of all points (x, y) of X with $x = 1$, a "vertical line". This subset of X is connected.

In the metric space of the last figure on page 49, the only connected subsets are those consisting of just one point, and the empty subset.

We turn, finally, to the definition of "bounded" for a set in a metric space. We begin by nothing that our old definition of "bounded", for a set in the plane, can indeed be immediately generalized to a subset of any metric space. It turns out, however, that the resulting definition is not a very convenient one. The reason for this is a rather technical one: One would have to look in detail at various possible applications of this notion of "bounded", and see that it just does not seem to fit properly. Perhaps one example will at least illustrate the idea. Consider the metric space X, d with X the plane, and $d(p, q)$ = geometrical distance between p and q if this is less than one, and one otherwise. We have see that this metric space is "very much like the plane", in that it always gives the same boundaries, the same interiors, and the same connected sets as the plane with usual geometrical distance. Does it give the same bounded sets? Let us try it, using the old definition of bounded (for X, d a metric space, and $A \subset X$, A is bounded if for some point p of X and some positive number c, every point q of A satisfies $d(p, q) \leq c$). Let, for example, $A = X$. Then, if our metric space is the plane with the usual geometrical distance, then A of course will not be bounded. Will A

be bounded in this metric space? The answer is yes. Let p be any point of X (say, $(-13, 233)$), and let $c = 2$. The locus of points within distance (using the "funny" d of this metric space) 2 of p is the entire set X. So, every point of A is within 2 of p. So, this A is bounded. Thus, if we carry over directly our old definition of bounded, we will get different answer for what is bounded in the plane with usual geometrical distance and the plane with this distance d.

What is the mechanism for this discrepancy? It is, roughly speaking, the following. There is only a difference between our distance d and usual geometrical distance for "larger distances" (namely, those exceeding one). On the other hand, all that really counts for boundaries, interiors, and connectedness is the "smaller distances". It was because geometrical distance and distance d agree for the "smaller distances" that these two metric spaces yielded the same boundary, interior, etc. On the other hand, what counts for the present definition of "bounded" is, not the "smaller distances", but rather the "larger distances", for, to show boundedness, one wants to choose c very large so that all of A will be within distance c of p. What we would really like, then, is some definition of "bounded" which concerns itself more with "smaller distances". An appropriate definition, as it turns out, is the following.

<u>Definition.</u> Let X, d be a metric space, and let $A \subset X$. Then A is said to be *bounded* if, for any positive number ϵ, there is a finite set of points of X, $p_1, p_2 \ldots p_n$, such that, for any point q of A, there is at least one of the p_i's such that $d(p_i, q) \leq \epsilon$.

What the definition requires, in other words, is that, given *any* positive ϵ, one can find a *finite* set of points in X such that every point of A is within ϵ of *at least one* of those points.

It is all very nice to write down a definition – but we had better check that it has at least some correspondence with our intuitive ideas of "bounded". The most convincing "check" would be to demonstrated that this definition yields exactly the same answers as our earlier definition, at least for the usual plane. The fact is that in this case, it does.

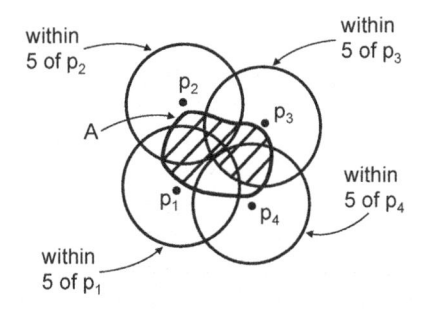

Let X, d be the usual plane. Let A be a subset of the plane, and suppose that A is bounded by the definition above. We shall show that A is bounded by the definition on page 17, i.e., we shall find a point p of X and a positive number c such that every point of A is within c of p. Since A is to be bounded by the definition above, we know that for every positive ϵ, there is a finite set of points of $X \ldots$ In particular, then, this must be true for $\epsilon = 5$. Thus, we have some finite set of points of X, $p_1 \ldots, p_n$, such that every point of A within 5 of at least one of the p_i.. Let us choose $p = p_1$. [Any of the p_i would do.] Then we can compute the distances $d(p, p_2)$, $d(p, p_3)$, and so on to $d(p, p_n)$. These will of course all be just real numbers. Let b denote the *largest* of those real numbers. What we have, then, is that all of $p_1, p_2 \ldots, p_n$ are within distance b of p ($= p_1$), since b was the largest distance. But every point of A is within distance 5 of one of the p_i. It now follows from the triangle inequality that every point of A is within distance $b + 5$ of p (for given any point q of A, there is some p_i with $d(p_i, q) \le 5$, while $d(p, p_i) \le b$, and so, by triangle, $d(p, q) \le b + 5$). Thus, we can set $c = b + 5$: Every point of A is within distance c of p. In short, we have shown that A is bounded by the definition on page 17.

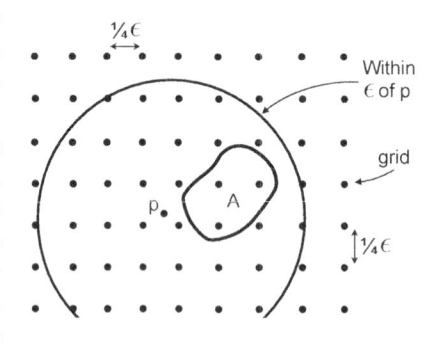

So, every set in the usual plane bounded according to the definition on the previous page is bounded according to the definition on page 17. We now show the reverse (which is a little bit trickier). Let A be a set in the plane bounded according to the definition on page 17. So, we have a point p and a positive number c such that every point of A is within c of p. We must show boundedness according to the definition on the previous page. Let any positive number ϵ be given. We must find a finite set of points of X such that every point of A is within ϵ of one of them. The method is the following. We introduce a "grid" of points in the plane, consisting of all points (x, y) of the form $n\epsilon/4$, $= m\epsilon/4$), where n and m are integers. This "grid" includes, for example, the points $(13\epsilon/4, -\epsilon/4)$ and $(-7\epsilon/4, 0)$ – it is a rectangular array of points in the plane, spaced $\epsilon/4$ apart. The point of this grid is that, because of the spacing between the points, *every point of the plane* is within distance ϵ of at least one of them. But, unfortunately, this grid will not do the job (for the $p_1, \ldots, p_n$) for the definition on page 63, because there are of course an infinite number of points in the grid. So, we do the following. We consider only the points of our grid which are within distance c of p, i.e., which are within the disk in the figure. This will be a *finite* number of points: call them $p_1, p_2, \ldots, p_n$. [Of course, "n" may be very large, perhaps in the millions, depending on how small ϵ is.] Now clearly every point in our disk, i.e., every point within distance c of p, is within distance ϵ of at least one of the $p_1 \ldots, p_n$ (for every point of the whole plane is within ϵ of *some* point of the original grid, and we obtained the p_i by just throwing away those points of the grid which do not make any difference as far as points in our disk are concerned). But every point of A is within distance c of p. So, we conclude that every point of A is within distance ϵ of at least one of the points $p_1, p_2, \ldots, p_n$. Thus, we began with a set A bounded according to the definition on page 17, chose any positive ϵ, and found a finite set of points, $p_1, p_2, \ldots, p_n$, such that every points of A is within ϵ of at least one of the p_i. We have shown, then, that every set bounded according to the definition on page 17 is bounded according to the definition on page 63.

Putting these together, then, we have shown (although we have not actually written out the formal proof for)

Theorem . Let X, d be the metric space which is the plane with usual geometrical distance, and let $A \subset X$. Then A is bounded according the definition on page 17 if and only if A is bounded according to the definition on page 63.

I hope the discussion above makes it clear what the mechanism of the definition on page 63 is. In order to show that a set is bounded, one must give oneself a positive ϵ, and find a finite number of points of X such that every point of A is within ϵ of at least one of those points. Generally speaking, "the smaller ϵ is, the more points that will be required in the finite set". Note that our new definition of "bounded" deals essentially with "the smaller distances".

The above is an example of a common practice in mathematics. One begins with a definition which seems "reasonable" in a certain context. (Here, "bounded" for sets in the usual plane). One broadens the context (here, to metric space), and then discovers that the old definition does not seem to work quite right in that broader context. So, one looks for a new definition (here, that on page 63) which essentially agrees with the old one in the context in which the old one was applicable (here, the new definition not only "essentially agrees", but "exactly agrees"), but which seems more appropriate in the broader context.

Hereafter, "bounded", for a set in a metric space, always refers to the definition on page 63.

We invented this new definition of "bounded" in part because of a certain example. Let us now go back and see what the new definition does to that example. Let X be the plane, and let $d(p, q)$ be the geometrical distance if it is less than one, and one otherwise. Set $A = X$. Is A bounded (according, of course, to the definition on page 63)? Let us try $\epsilon = 3$. Is there a finite set of points of X such that every point of A (i.e., every point of X) is within 3 of at least one of them? There is: Let p_1, any point be the "finite set of points". The locus of points within 3 of p_1 is all of X, and no every point of A is within 3 of "one of them" (namely, the only one, p_1). But, to show bounded, we have to do this for *every* positive ϵ. What about $\epsilon = 1/2$? Is there a finite set of points in X such that every point of A is within $1/2$ of at least one of them? The locus of points within $1/2$ of a point is a disk of radius $1/2$ centered at that point. Can we find a *finite* number of such disks (all of radius $1/2$) such that every point of X ($= A$) is in at least one of those disks? Obviously, we cannot. So, we cannot find such a finite set for $\epsilon = 1/2$. So, A in this example is not bounded.

Note how clever the new definition is. It "corrects" the answer in this example, to yield that the entire set X is not bounded. It does so in a way so that no answers are "corrected" for the usual plane with geometrical distance, and also such that "small distances are the ones which really count". Let X be the plane, and let $d(p, q) = 1$ if $p \neq q$, and zero otherwise. Is $A = X$ bounded? Try $\epsilon = 1/2$. Is there a finite set $p_1 \ldots, p_n$ in X such that every point of X ($= A$) is within $1/2$ of at least one of the p_i? The locus of points within $1/2$ of a point p consists of just p itself. So, if every point of X is to be within $1/2$ of one of the p_i, every point of X must *be* one of the p_i. But this is impossible, for one is allowed only a finite number of p_i, while X itself has an infinite number of points. So, X itself is not bounded, since we cannot find such a finite set in X for $\epsilon = 1/2$. Clearly, the only subsets of X which are bounded are the finite subsets of X. [Of course, every finite subset of X is bounded. If A, consisting of just points $p_1 \ldots, p_n$, is a finite subset of X, and positive ϵ is given, consider the finite subset of X consisting of A. Then every point of A will be within ϵ of at least one point of this finite set.]

Let X be the set of all disks in the plane, with d given using areas, as before. For $A = X$, A is not bounded. [Try $\epsilon = 1$. Can one find a finite set of disks such that every disk is within distance (using areas) 1 of one of these?] The subset of X consisting of all disks with center the origin is also not bounded. The subset of X consisting of all disks with center the origin is also not bounded. The subset consisting of all discks which lie within a square is bounded (but it is a bit tricky to see).

For the metric space of the last figure on page 49, every subset of X is bounded.

Let X be the points (x, y) of the plane with x rational. For A the subset of X consisting of all (x, y) with $x^2 + y^2 < 1$ and x rational, A is bounded; for A consisting of all (x, y) with $x = 1$, A is not bounded. For $A = X$, A is not bounded.

8. Theorems on Sets in Metric Space

We now have the notion of a metric space, and some examples; the definitions of interior, boundary, connected, and bounded in a metric space, and some examples. The next step is the same as the corresponding next step for the plane: We wish to prove some theorems which relate these definitions to each other. One expects, of course, that many of our old theorems for the plane will have analogous versions for metric spaces (with only "typographical changes", such as "point of the plane" becoming "point of X", etc). This feature, after all, was one of our guides in inventing metric spaces. In fact, one often thinks up theorems and their proofs in metric spaces by "thinking plane" (or some other very simple metric space), but "writing metric space". One must be a bit careful in doing this, however, for, when proving a theorem about metric spaces, one is of course only allowed to use those properties of distance which are given in the definition of a metric space. It even turns out on occasion that it is easier to prove something in the more general context of a metric space than in the more special context of the plane. Whereas there are millions of properties of geometrical distance in the plane, there are only three properties of distance in a general metric space. Thus, in the case of metric spaces, there are fewer possibilities for what to do if, in the middle of a proof, one is trying to get from something to something.

Below are various theorems in metric spaces. Most generalize earlier results for the plane. We occasionally omit proofs which are virtually identical to earlier proofs.

Theorem. Let X, d be a metric space, and $A \subset X$. Then int $(A) \subset A$. Proof: Let p be a point of int (A), so, for some positive number ϵ, every point within ϵ of p is in A. But p is within ϵ of p, and so p is in A.

[This is of course the theorem on page 26, but now for metric spaces. The "every point" of the first sentence in the proof means "every point of X", and "within" means "within using distance d of the metric space". That p is within ϵ of p follows from the first condition in the definition of a metric space.]

Theorem. Let X, d be a metric space, and let $A \subset X$. Then int (int (A)) = int (A).

Proof: By the theorem above, every point of int (int (A)) is in int (A). For the converse, let p be a point of int (A). Then for some positive ϵ, every point within ϵ of p is in A. Let q be any point within $1/2\epsilon$ of p. Then every point within $1/2\epsilon$ of q must, by the triangle inequality, be within ϵ of p, and so must be in A. So, q must be in int (A). Since every point within $1/2\epsilon$ of p is in int (A), p is in int (int (A)).
[See theorem on page 28. All we added was "by the triangle inequality".]
Theorem. Let X, d be a metric space, and $A \subset B \subset X$. Then int $(A) \subset$ int (B).

[See the first theorem on page 30.]
Theorem. Let X, d be a metric space, and $A \subset X$ and $B \subset X$. Then int $(A \cap B) =$ int $(A) \cap$ int (B).
[See the second theorem on page 30.]
Theorem. Let X, d be a metric space, and $A \subset X$ and $B \subset X$. Then int $(A) \cup$ int $(B) \subset$ int $(A \cup B)$.
[See theorem on page 32. The proof is that, word or word.]
Theorem Let X, d be a metric space, and $A \subset X$. Then bnd $A^C) =$ bnd (A).
[See theorem on page 29.] This, recall, was one of our "interesting" theorems for the plane, for it was suggested by our intuitive ideas of "boundary" and of sets in the plane. There are some pretty strange things which still get to be called metric spaces. Yet, this theorem holds for them all.]
Theorem. Let X, d be a metric space, and $A \subset X$. Then every point of X is in either int A or bnd (A) or int (A^C), and no point of X is in more than one of these sets.
Proof; Let p be a point of X. If for some positive number ϵ every point within ϵ of p is in A, then p is in int (A). If for some positive number ϵ every point within ϵ of p is not in A, then p is in int (A^C). Clearly, p cannot be in both int (A) and int (A^C). If neither of these holds, i.e., if for every positive number ϵ there is a point within ϵ of p and in A and also a point within ϵ of p and not in A, then p is in bnd (A). Finally, if p is in bnd (A), then (since for every positive ϵ there is a point within ϵ of p and not in A) p is not in int A, and (since for every positive ϵ there is a point within ϵ of p and in A) p is not in int (A^C).
[We did not actually state this as a theorem for the plane, although of course we could have. Note that the second sentence of the theorem could have been "Then int $(A) \cup$ bnd $(A) \cup$ int (A^C) is the set X, and int $(A) \cap$ bnd (A), bnd $(A) \cap$ int (A^C), and int $(A) \cap$ int (A^C) are all the empty set." In the proof, one shows in succession: p can be in int (A); p can be in int (A^C); p cannot be in both; if p is in neither, then it is in bnd (A); if p is in bnd (A), then it can be in neither int (A) nor int (A^C). These assertions, taken together, establish the theorem.]
Theorem. Let X, d be a metric space, and $A \subset B \subset X$. Then, if B is bounded, so is A.

Proof: Let ϵ be a positive number. Then, since B is bounded, there is a finite set, $p_1 \ldots, p_n$ of points of X such that every point of B is within ϵ of one of them. But $A \subset B$, and so every point of A is within ϵ of one of these points. So, A is bounded. [See theorem on page 32. Of course, now we must use the "corrects" definition of bounded, (page 63), and so the present proof is quite different from the earlier proof for the plane. It is reassuring that, even with our new definition of "bounded", we still get this theorem. Note how the present proof begins. We are, after all, trying to show that A is bounded, i.e., that "given any positive $\epsilon, \ldots$". So, we start by giving ourselves a positive ϵ, and we finish by finding a finite set such that every point of A is within ϵ of at least one of the points of that set.]

Theorem. Let X, d be a metric space, and let A and B be bounded subsets of X. Then $A \cup B$ is bounded.

Proof: Let ϵ be a positive number. Since A is bounded, there is a finite set, $p_1 \ldots, p_n$, such that every point of A is within ϵ of one of these points: since B is bounded, there is a finite set, $q_1 \ldots, q_m$, such that every point of B is within ϵ of one of these points. Consider the finite set $p_1 \ldots, p_n, q_1, \ldots, q_m$. Then every point of $A \cup B$, since it must be in either A or B, must be within ϵ of one of these points. So, $A \cup B$ is bounded. [Again, we did not state this one as a theorem for the plane. The present proof is actually a bit simpler than would be the corresponding proof using the old definition of "bounded" in the plane. Of course, one would have expected this result from one's intuitive idea of "bounded". (Score one for the definition.) How does one think of such a proof? One knows that A and B are bounded, so one thinks of a figure such as that on the right, with a "grid" of points such that every point of A is within ϵ of one of them, and similarly for B. One wants to show that $A \cup B$ is bounded, i.e., one must find a new "grid" such that every point of $A \cup B$ is within ϵ of one of these points. What should one choose for the new "grid"? Clearly, one should just combine the grids for A and B. Note that the proof begins "Let ϵ be a positive number."]

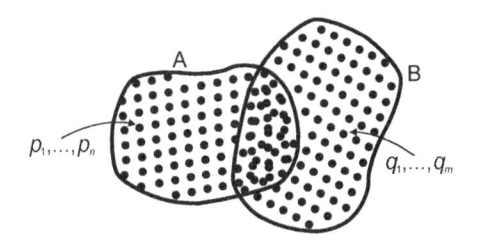

Theorem. Let X, d be a metric space, and let A and B be connected subsets of X having some point p in common. Then $A \cup B$ is connected. [The proof is, word for word, that of the theorem on page 33. Again, we have a result which, at least for the plane, reflects our intuitive idea of what "connected" means. But, despite the fact that there are some very strange metric spaces around, it even holds in a general metric space. That theorems such as this one continue to hold gives one confidence that the introduction of a metric space was the right idea.]

Finally, we remark that, since the usual plane is of course a metric space, all our earlier counterexamples for the plane remain counterexamples for metric spaces. Thus, for example, the following statements (in which "..." stands for "Let X, d be a metric space, and $A \subset X$ and (if B is mentioned) $B \subset X$ are *false*: i) ... No point of A is in bnd (A). ii) ... int (bnd (A)) is the empty set. iii) ... If $A \subset B$, then bnd $(A) \subset$ bnd (B). iv) ... int (bnd $(A \cup B)$ = int $(A) \cup$ int (B). v) ... If A and B are connected, so is $A \cap B$. vi) ... If $A \subset B$ and B is connected, then A is connected. vii) ... If A is connected, then A^c is connected. viii) ... If A is bounded and $A \subset B$ then B is bounded. ix) ... If A is bounded, then A^c is bounded.

9. Topological Space

We are now at another crossroads. We have the notion of a metric space, some definitions in metric spaces, some examples, and some theorems which relate the definitions to each other. What should be done next? Shall we invent more definitions, more examples, and more theorems? This would certainly be possible. However, just with the plane before, this is not the direction in which we wish to proceed.

We begin with the observation that there is still something fishy about all this. Recall our discussion of stretching and pulling of the rubber sheet. We decided initially that the kinds of things we were after were those which remained the same under this operation of stretching and pulling. We indeed found four such "things": the interior and boundary, and the notions of connected and bounded. We asked what was genuinely relevant to these notions, and decided that it was distance. This observation was made manifest by expressing interior, boundary, connected, and bounded in a general metric space (i.e., in "a space in which the only thing one has access to is a distance"). What is disquieting, however, is the following fact: Whereas interior, etc. certainly seem to remain the same under the operation of stretching and pulling, "distance" certainly does not. That is to say, the operation of stretching a rubber sheet will certainly change distances between points of the sheet. How does it come to be that, in order to express "stretching-independent" notions, we need something "starching-dependent", namely distance? A more explicit example of this phenomenon is a result from Sect. 7. Let X be the plane, d' the usual geometrical distance, and d the distance with $d(p, q) = d'(p.q)$ if $d'(p, q) < 1$, and $d(p, q) = 1$ otherwise. Then, as we have seen, for any subset A of X one gets the same interior and boundary whether one uses d or d'; the same sets are bounded or connected whether one uses d or d'. These two distances are of course quite different, whereas as far as the structural features we are interested in are concerned, they "might as well be the same". It is as though d and d' different from each other only by a stretching and pulling of the metric space.

The mathematician would regard it as unpleasant that, in order to treat certain features of sets in the plane, one has to invoke structure (namely, a distance) which seems to go beyond what is inherent in the features one is after. It would be taken as the sign that one does not yet understand that one is doing – that one has not yet isolated what is genuinely relevant to "interior", etc. One could hardly doubt that distance has *something* to do with our notions. It is just that it does not seem to be the essential thing – the thing which gives us access to our four notions without forcing one to carry around excess baggage. As the example above makes clear, there is certainly a great deal of "excess baggage" in distance.

So, what is the essential thing? What is *really* relevant here? Here is a problem in mathematics. One might begin, for example, by looking back at what we have done for some clues to the answer. One might in this way stumble across the following clues. First, one might notice that we tend to use our distances in a rather restrictive way. For instance, there never appears, in anything important, a statement such as "The distance is 17.". In fact, the only circumstance in which we actually use our distances is the following. We have some point p. We then say either "for some positive number ϵ" or "for all positive numbers ϵ, and then either "there exists point q with $d(p, q) \leq \epsilon$" or "for all points q with $d)p, q) \leq \epsilon$". It seems as though we do no care about the exact, specific numerical value of $d(p, q)$, but only whether or not it is $\geq \epsilon$, and then, not for a specific numerical value for ϵ, but rather for either "all ϵ "some ϵ". A second clue is an observation we made, and even used, in Sect 7. What seems to count about our distances is, not the larger distances, but only the smaller distances. Thus, for the example on the previous page, we had two distances which "agreed when they are small, but not when they are large". Theses two distances, however, gave rise to the same interior, etc. But even or the smaller distances, it is not exactly the numerical values which count. For example, let X be the plane, d' usual geometrical distance, and d twice geometrical distance. These distances are of course different – and they are even different for "small distances". However, one easily convinces oneself that d and d' give rise to the same interiors and boundaries, and to the same connected sets and bounded sets.

These clues describe different aspects about what it is we want to retain of our distances. It is "the smaller distances, but even then not exactly their numerical values, but rather their '$\leq \epsilon$ for some ϵ' and '$\leq \epsilon$ for all ϵ' structure". What we want to retain is a sort of shadow of distance which reflects these features.

But the most revealing clue is the following. Consider the fifth theorem on page 68. It says that, given any set A in any metric space, each point of the space is in exactly one of the sets int (A), bnd (A), and int (A^c). But this means the following. Suppose that one were very good at determining interiors, but rather poor at boundaries. Then one could use interiors to determine boundaries. Indeed, because of this theorem, one could determine bnd (A) by finding all points which remain after removing int (A) and int (A^c) from

X. Now consider the definition of "disconnected" on page 60. Where does distance enter this definition? At only one place: the "... bnd (B)." at the end (for, of course, distances are needed to determine the boundary). That is to say, if one has determined the boundaries of all sets, then even if one has lost one's sheet giving the distances, one can still determine which sets are disconnected and which are not. We conclude, then, that if one knows the interiors then one can determine the boundaries (without otherwise using the distance), and further that, once one has determined the boundaries, one can determine the disconnected sets (again, without otherwise using the distance). [What of "bounded"? It does not seem to fit into this scheme. We shall return to this shortly.] In short, distance seems to enter this subject essentially only through the determination of interiors.

Here, then, are the clues. The mystery is how to so formulate this subject that the essential things are retained, while the excess baggage is eliminated. Of course, we have very much prejudiced the issue by our choice of clues. In practice, one might find eighty clues, which point in twelve different directions. One would begin with the most promising of these directions, and work one's way along, hoping eventually to find a formulation which seems to have the appropriate character.

The solution to our mystery is the following. What did we do when we passed from the plane to a general metric space? For the plane, we *computed* the distance, using the formula on page 10.

For metric spaces, however, distances were not *computed*: rather, they were just *specified* (i.e., invented), as part of the definition of a metric space. We now want to pass from a metric space to something else. The crucial thing seems to be the interiors. For a metric space, the interiors are *computed* (using the distance in that metric space, and the definition of "interior"). The idea, then, is that in our new spaces, the "interiors will just be specified by fiat, and not computed from anything else". In this way, we shall get rid of those awkward distances, but retain the essential thing we need about the distance, namely the interiors it determines.

But how can one tell, in a metric space, if a given set is the interior of something? That is easy.

<u>Theorem</u>. Let X, d be a metric space, and $A \subset X$. Then there exists a subset B of X such that $A = \text{int}(B)$ if and only if $A = \text{int}(A)$.

Proof: If $A = \text{int}(B)$ for some $B \subset X$, then, by the last theorem on page 67, $\text{int}(A) = A$. For the converse, let $A = \text{int}(A)$. Then there is a set B (namely, A) such that $A = \text{int}(B)$.

Thus, the sets which are the interiors of *something* are just the sets which are the interiors of *themselves*.

<u>Definition</u>. Let X, d be a metric space, and $A \subset X$. Then A will be said to be *self-interior* if $A = \text{int}(A)$.

It is the self-interior sets which seem to play the critical role in the study of sets in a metric space.

Recall again what happened in the passage from the plane to the general metric space. We first isolated distance as the essential thing. We next obtained a list of the "distance-like" properties of distance. The final step was to assemble these properties in the definition of a metric space. For the present situation, the essential thing seems to be the self-interior sets. The next step, then, is to figure out what are the "self-interior-like" properties of self-interior sets. With the plane, we were looking for properties of distance which referred only to distance, and not to anything else about the plane. Similarly, we are here looking for properties of self-interior sets which refer only to these sets, and not to anything else about the metric space. Thus, for example, "For A self-interior, there is a set B such that $A = $ int (B)." is not appropriate, for it refers to "int", which requires the distance in our metric space. By contrast, "For A self-interior, A^c is self-interior." would be an acceptable property by this criterion, for it refers only to the self-interior sets (and, of course, to the complement, but that is ok, because one does not need the distance to determine the complement). But this property is not a very good one either, for it is false in a general metric space. [For example, let the metric space be the usual plane, and A the usual disk. Then A is self-interior, but A^c is not.] Thus, we are looking for properties which refer only to self-interior, and which are true in a general metric space. Again, there are many more properties than we shall want to retain in the final definition (just as there are many many more properties of distance in the plane than are retained in the definition of a metric space). For example: "For p and q points of X, with $p \neq q$, there is a self-interior set A with p in A and q not in A." is true in any metric space. We shall not retain it. Again, it is a matter of judgment, to select enough properties to capture the essence of what one wants, but not so many that one "practically gets all the way back to a metric space".

The appropriate list of properties, as it turns out, is the following.

Theorem. Let X, d be a metric space. Then

1. X and the empty subset of X are self-interior.

2. For A and B self-interior, $A \cap B$ is self-interior.

3. Given any collection of self-interior sets, their union C is self-interior.

Proof: 1. For p in X, every point within distance 13 of p is in X, and so X is self-interior. The empty set, since it contains no points, is self-interior. 2. Let A and B be self-interior. Then int $(A \cap B) = $ int $(A) \cap $ int $(B) = A \cap B$, where the first equality follows from the first theorem on page 68, and the second equality follows from $A = $ int (A) and $B = $ int (B). 3. By the second theorem on page 67, int $(C) \subset C$. For the converse, let p be a point of C. Then p must be in one of the self-interior sets of the union, say A. Since $A = $ int (A), p is in int (A), i.e., there is a positive number ϵ of p is in A. But $A \subset C$, and so every point within ϵ of p is in C. So, P is in int (C). We conclude that C is self-interior.

Note that, in the third property, we show that the union of *any* collection (possibly two, possibly nineteen, possibly an infinite number) of self-interior sets is self-interior. Of course, the union of an arbitrary collection of sets is the set of all points which are in any one of those sets; the intersection of an arbitrary collection of sets is the set of all points which are in all of those sets. For example, let X be the plane, and consider the collection of subsets of X consisting of all disks (of any positive radius) centered at the origin. That is, for a any positive number, the subset of X consisting of all (x,y) with $x^2 + y^2 < a$ is in our collection. Then the union of this collection is the entire plane (for every point of the plane is in at least one of these disks.) The intersection of this collection is the set consisting of the single point at the origin (for this is the only point which is in every one of our disks). In this example, all the sets in our collection are self-interior in the metric space with d usual geometrical distance. Note that their union, the entire plane, is also self-interior, as demanded by the third property. Their intersection, however, is not self-interior. This example shows, then, that in the second property of the theorem above, we could not have had that the intersection of any collection of self-interior sets is self-interior. Thus, for self-interior sets, one can take intersections of two, but unions of arbitrary numbers, and still get self-interior sets.

The next step is obvious. Just as the decision as to what are the "appropriate" properties of distance in the plane led to metric spaces, so the decision as to the appropriate properties of self-interior sets leads to a kind of space.

Definition. A *topological space* consists of a set X together with a collection $\mathcal{T}$ of subset of X such that the following conditions are satisfied:

1. X and the empty subset of X are in the collection $\mathcal{T}$.
2. For any sets A and B in the collection $\mathcal{T}$, $A \cap B$ is in the collection $\mathcal{T}$.
3. For any collection of sets in $\mathcal{T}$, their union is also in $\mathcal{T}$.

Thus, to specify a topological space, one must say what the set X is, and what the collection $\mathcal{T}$ of subsets of X is (i.e., which subsets of X are to be in the collection $\mathcal{T}$, and which are not). The subsets in the collection $\mathcal{T}$ are usually called *open* sets (in the topological space X, $\mathcal{T}$), i.e., "open set" is just a short way of saying "subset of X which is in the collection $\mathcal{T}$".

The definition of of a topological space is, in my opinion, one of the most beautiful definitions in mathematics. On the one hand, the definition itself is simple and elegant. [One only has to specify a set X and some subsets, subject to simple conditions. One does not have to give a lot of complicated, detailed information, such as the distances for a metric space.] On the other hand, this single definition gives one access to an enormous and rich range of intuitive ideas – those we have discussed so far as well as many others. The study of topological spaces is a large and important branch of mathematics –

called, of course, topology. This will be the subject of our attention hereafter. I hope that we shall be able to give some insight into the power and elegance of this definition.

We give some examples of topological spaces. Just as the plane led immediately to an example of a metric space, so any metric space leads immediately to an example of a topological space. Indeed,

Theorem. Let X, d be a metric space. Denote by $\mathcal{J}$ the collection of all self-interior (using d) subset of X. Then $X, \mathcal{J}$ is a topological space.

Proof: This is immediate from the theorem on 74. Note that, given metric space X, d, we specified the information we needed to specify to have a candidate for a topological space, namely a set (here, X), and a collection of subsets of X (here, all self-interior subsets, using d). We already have the necessary properties, since after all it was just these properties for self-interior subsets of a metric space which suggested our definition of a topological space. The topological spaces so constructed are important enough to deserve a name.

Definition. Let X, d be a metric space. The topological space $X, \mathcal{J}$ obtained by the theorem above will be called the *underlying* topological space of the metric space X, d.

Thus, we instantly obtain many topological spaces, as the underlying topological spaces of our various metric spaces. For example, let X be the plane, and d usual geometrical distance. Then, for the underlying topological space of this metric space, X is of course the set of points of the plane, while $\mathcal{J}$ (the collection of open sets of our topological space) is the collection of self-interior sets of the plane (e.g., the empty set, X itself, any disk, any union of disks, and so on). Next, let X be the plane, but let d be the distance which is geometrical distance if that is less than one, and one otherwise. Then, as we have seen, X, d is a metric space. What is the underlying topological space? The set is of course X, the plane. Now, however, the open sets are to be those self-interior using this d. What, are these sets? Recall that, on page 68, we showed that the interior, using *this* distance d, is always exactly the same as the interior using the usual geometrical distance. Clearly, then, the self-interior sets, in *this* metric space, will be exactly the same as the self-interior sets using the geometrical d. Thus, we obtain exactly the same underlying topological space as we obtained above, for the usual distance in the plane. [How nice! One of the things we were unhappy about was that we seemed, with metric space, to be carrying around excess baggage, reflected by the fact that this distance and geometrical distance seemed to be "essentially the same as far as the things we are interested in are concerned". The definition of the underlying topological space automatically throws away the excess baggage. We get the same underlying topological space in the two cases.]

Let X be the plane, and let d be given by $d(p, q) = 0$ if $p = q$, and $d(p, q) = 1$ if $p \neq q$. This is a metric space. The set of its underlying topological space is just the plane X. To find the open sets of this topological space, we must find the self-interior sets of our metric space. When, in this metric space, is int $(A) = A$? This is *always* true, for any $A \subset X$, in this metric space. Hence, every subset of X is an open set in this topological space. More generally, we have

Theorem. Let X be a set, and let $\mathcal{J}$ be the collection of all subsets of X. Then X, $\mathcal{J}$ is a topological space.

Proof: 1. X itself and the empty subset are subsets of X, hence in $\mathcal{J}$. 2. The interaction of two subsets of X is a subset of X, hence in $\mathcal{J}$. 3. The union of any collection of subsets of X is a subset of X, hence in $\mathcal{J}$.

Thus, for *any* set X we obtain, via this theorem, a topological space (namely, one in which every subset of X is open). These have a special name.

Definition. A topological space X, $\mathcal{J}$ for which every subset of x is in $\mathcal{J}$ is said to be *discrete*.

Consider the metric space on page 49, in which X consists of just four points, with distances 17 and 1. Its underlying topological space (in which the X, of course, will be just this set with four points) is discrete.

Of course, to give a topological space, one must give nothing less, and need give nothing more, than required by the definition. Every metric space, as we have just seen, gives rise to a topological space – but this need not be the *only* way of obtaining topological spaces. Here is another kind of topological space.

Theorem. Let X be a set, and let $\mathcal{J}$ be the collection of subsets of X consisting of X itself and the empty subset. Then X, $\mathcal{J}$ is a topological space.

Proof: 1. X and the empty set (here written $\emptyset$) are in $\mathcal{J}$. 2. Since $X \cap X = X$, $X \cap \emptyset = \emptyset$, $\emptyset \cap X = \emptyset$, and $\emptyset \cap \emptyset = \emptyset$, the intersection of any two sets in $\mathcal{J}$ is in $\mathcal{J}$. 3. The union of any collection of sets in $\mathcal{J}$ is X if any one of those sets in the union is X, and $\emptyset$ if none of those sets in the union is X. So, any such union is in $\mathcal{J}$.

The topological spaces obtained via this theorem have "as few open sets as possible", namely just X and the empty set. [Clearly, one cannot have have fewer open sets than this, because one has to satisfy the first condition for a topological space.] By contrast, a discrete topological space has "as many open sets as possible", namely every subset of X is open. These two, then, are the extremes for how many open sets there can be. The one with "very few open sets" also has a name.

Definition. A topological space X, $\mathcal{J}$ for which $\mathcal{J}$ consists only of X itself and the empty subset is said to be *indiscrete*.

Just as, in a metric space, "X does not tell one what distance to choose", so, in a topological space, "X does not tell one what open sets to choose". One's only obligation is to specify clearly what subsets of X are or are not in, $\mathcal{J}$, and check the conditions.

Let X be the plane, and let $\mathcal{J}$ consist of all subsets of X having a finite number of points (including the possibility of no points). Thus, the usual disk is not in $\mathcal{J}$, for it has more than a finite number of points. We check the conditions for a topological space. The empty subset of X is certainly in $\mathcal{J}$, for it has no points. However, X itself is not in $\mathcal{J}$, for X does not have a finite number of points. The first condition fails. For the second condition: Is it true that, if each of A and B has a finite number of points, then $A \cap B$ has a finite number of points? This is true, and so the second condition holds. For the third condition: Is it true that, given any collection of subsets of X, each with a finite number of points, their union has a finite number of points? Let us try it for two subsets in the collection. Is it true that, if each of A and B has a finite number of points, then $A \cup B$ has a finite number of points? This is true. However, this third condition fails in general (i.e., when we allow ourselves to take the union of *any* collection of sets in $\mathcal{J}$). Indeed, consider the collection of all subsets of X having just one point (so the subset consisting of just the origin is in this collection, while a disk is not). This is certainly a collection of sets in $\mathcal{J}$ (for each set in our collection has a finite number of points, namely just one point). What is the union of this collection of sets? It is, we claim, the entire plane X. [Indeed, let p be any point of the plane. Then the subset of X consisting of just p is in our collection. So, p is in the union of the sets in our collection.] But the plane, of course, is not in $\mathcal{J}$. So, we have found a collection of sets in $\mathcal{J}$ whose union is not in $\mathcal{J}$. The third condition fails. Thus, we do not have a topological space (and, in fact, we only satisfied one condition out of the three).

The "mirror image" of the attempt above, however, works. Let X be the plane, and let $\mathcal{J}$ consist of the empty set, X itself, and any subset of X consisting of all of X except for a finite number of points. Thus, for example, the disk is not in $\mathcal{J}$, for it is not the empty set, not X, and it is not all of X except for a finite number of points. But the subset of X consisting of all (x, y) except $(3, -271)$ is in $\mathcal{J}$, for it is all of X except for a finite number of points (namely, except for one point). This X, $\mathcal{J}$, we claim, is a topological space. 1. X and the empty set are certainly in $\mathcal{J}$. 2. Let A and B be in $\mathcal{J}$. If either is the empty set, then $A \cap B$ is the empty set, and so is in $\mathcal{J}$. Suppose, then, that neither is the empty set: Say, A is all of X except the finite number of points $p_1 \ldots, p_n$ of X, and B is all of X except the finite number of points $q_1 \ldots, q_m$ of X. Then $A \cap B$ (the set of all points in both A and B) must be all of X except for points $p_1 \ldots, p_n, q_1 \ldots, q_m$ (for every other point of X is certainly in both A and B, while these points are not in both A and B). But this set $A \cap B$ is all of X except for a finite number of points, and so is in $\mathcal{J}$. 3. Consider any collection of sets in $\mathcal{J}$, and let C be their union. If all the sets in this union are the empty set, then C is the empty set, and so C is in $\mathcal{J}$. Suppose not, so there is some set A in this union which is not the empty set. Then A, since it is in $\mathcal{J}$, must be all of X except for a finite number of points.

But C is a union of A with some other sets, and so $A \subset C$. But, since A is all of X except for a finite number of points, and $A \subset C$, C must be all of X except for a finite number of point (for C has to include at least everything in A). So, C is in the collection $\mathcal{J}$. That is, the union of any collection of sets in $\mathcal{J}$ is itself in $\mathcal{J}$. We conclude, therefore, that this is a topological space.

Compare this last example (in which we obtain a topological space) with the preceding one (in which we do not). Why is "finite subsets' of X" so different from "subsets of X consisting of all of X accept for a finite number of points"? One easily traces the reason to the difference between the second and third conditions for a topological space: Whereas the intersection of just *two* open sets must be open, the union of *any collection* of open sets must be open.

Let X consist of three points, p, q, and r. Let $\mathcal{J}$ consist of the empty subset of X, X itself, the subset of X consisting of just p, and the subset of X consisting of just q and r. One checks that this $X, \mathcal{J}$ is a topological space. If, however, $\mathcal{J}$ had consisted of the empty subset of X, X itself, the subset of X consisting of p and q, and the subset of X consisting of q and r, then we would not obtain a topological space: The second condition would fail.

We introduce one final method for obtaining topological space. Recall the following construction for metric spaces. Let X, d be a metric space, and let X' be *any* subset of X. We introduce a distance d' in the set X', as follows. Let p and q be any two points of X'. Then, since $X' \subset X$,, p and q are also points of X. But we already have a distance, d in X. We now *define* our distance in X' to be just this distance in X, i.e., we set $d'(p, q) = d(p, q)$. We indeed have specified a rule which assigns a number (namely, $d'(p, q')$) to any two points, p and q, of X'. We finally note that this X, d' is a metric space. [Each of the three conditions for X', d' to be a metric space follows immediately from the corresponding condition in the metric space X, d.] There is an analogous construction in topological spaces. Let $X\mathcal{J}$ be a topological space, and let $X' \subset X$. Just as above we used the distances in X to obtain distances in X', now we wish to use the open sets in X (i.e., those in the collection $\mathcal{J}$) to obtain our open sets X' (i.e., those which will be in our collection $\mathcal{J}'$). This we do as follows. Let A be any open set in the topological space $X, \mathcal{J}$. We cannot of course simply say "We shall then regard this A an an open set in $X', \mathcal{J}'$.", for A may not even be a subset of X'. We therefore "force" A to e a subset of X', as follows. We simply take the intersection of A (a subset of X) and X' (also a subset of X). This intersection (of course, a subset of X) will of course also be a subset of X'. It is these subsets of X' (namely, all those obtained by intersecting open sets in X, $\mathcal{J}$ with X') that we wish to deem open in X', $\mathcal{J}'$. With metric spaces, the fact that our X', d was a metric space was immediate from the fact that X, d was a metric space. Similarly, here, one shows that $X', \mathcal{J}'$, constructed as above, is a topological space using the fact that $X, \mathcal{J}$ began as a topological space.

That is,

Theorem. Let $X, \mathcal{J}$ be a topological space, and $X' \subset X$. Let $\mathcal{J}'$ denote the collection of all subsets of X' of the form $A \cap X'$, where A is in the collection $\mathcal{J}$. Then $X', \mathcal{J}'$ is a topological space.

Proof: 1. Since the empty set $\emptyset$ is in $\mathcal{J}$, and $\emptyset \cap X' = \emptyset$, the empty set is in $\mathcal{J}'$. Since X is in the collection $\mathcal{J}$, and $X \cap X' = X'$, the set X' is in $\mathcal{J}'$. 2. Let A' and B' be in $\mathcal{J}'$, say $A' = A \cap X'$ and $B' = B \cap X'$, with A and B in the collection $\mathcal{J}$. Then $A' \cap B' = (A \cap B) \cap X'$, while, since $X, \mathcal{J}$ is a topological space, $A \cap B$ is also in $\mathcal{J}$. Hence, $A' \cap B'$ is in $\mathcal{J}'$. 3. Let C' be the union of any collection of sets in $\mathcal{J}'$. Then each of the sets in this union is the intersection, with X', of a set in the collection $\mathcal{J}$. Denote by C the union of the corresponding sets in $\mathcal{J}$. Then, since $X, \mathcal{J}$ is a topological space, this C is also in the collection $\mathcal{J}$. But we have $C' = C \cap X$. Hence, C' is in $\mathcal{J}'$.

In this proof, we have used a few (easy) facts about intersections and unions. [For example, in the second step, we used the fact that if one takes the intersection of $A \cap X'$ with $B \cap X'$, the result is the same as taking the intersection of $A \cap B$ with X'. In the third step, we used the fact that the result of taking the union of a collection of sets of the form $A \cap X'$ is the same as the result of first taking the union of all the A's, and then intersecting the result with X'. One checks these facts by the usual, straightforward, arguments. (E.g., "Let p be a point of .... Then p must be in ... or ... but not .... Therefore, p must be in ... etc.")] Using this theorem, then, we again obtain an enormous variety of topological spaces. One begins with *any* topological space, $X, \mathcal{J}$. One takes *any* subset X' of X. Then, "restricting" the open sets of $X, \mathcal{J}$ to X' (i.e., taking their intersections with X'), we obtain a new topological space, $X', \mathcal{J}'$. These topological spaces have a special name.

Definition. Let $X, \mathcal{J}$ be a topological space, and $X' \subset X$. Then the topological space $X', \mathcal{J}'$, obtained via the theorem above, is called the *topological subspace* (of $X, \mathcal{J}$, based on the subset X').

To summarize, any set X gives rise to two topological spaces (the indiscrete one and the discrete one); any metric space gives rise to a topological space (the underlying one); any metric space gives rise to a topological space (the underlying one); any subset of any topological space rise to a topological space (the topological subspace).

10. Interior, etc. in Topological Spaces

We initially introduced the notion of the interior, boundary, connected, and bounded for sets in the plane. These notions satisfied certain properties. We then introduced a sort of "generalized plane", a metric space. We were able to recover, in metric spaces, our four notions, with similar properties. We have now introduced "generalized metric spaces", i.e., topological spaces. We now wish to recover our four notions again in a general topological space, and again find their properties.

In our discussion of sets in the plane, we constantly made use of "disks of various radii centered at various points". In metric spaces, the analogous things were the subsets of X given by "the locus of all points q of X with $d(p,q) < \epsilon$". Subsets of this form were used in virtually every argument for metric spaces. In particular, these subsets of a metric space reduce, for the case of the plane, to our usual disks. There is a similar class of subsets for a topological space, which subsets will play very much the same role. These are the following.

Definition. Let X, $\mathcal{T}$ be a topological space, and p a point of X. An *open neighborhood* of p is an open set in this topological space, containing the point p.

We intend to use open neighborhood in topological space very much as we used disks in the plane, and "the locus of points q with $d(p,q) < \epsilon$" in a metric space. Is this at all feasible? Do open neighborhoods bare any resemblance to "the locus of points q with $d(p,q) < \epsilon$"? We have to ask this question with a little bit of care, We cannot, for example, just go off and ask "Are the open neighborhoods just given by the locus of points $q\ldots$?". The problem here is that open neighborhoods are defined in *topological spaces*, while "the locus of points $q\ldots$" are defined in *metric spaces*. But topological spaces and metric spaces are quite different things. We do not have any distance in a topological space, and we do not have any open sets in a metric space. But there is, fortunately, one area of common ground, on which we can check things out. We know that every metric space gives rise to a topological space, its underlying topological space. Thus, we may begin with a metric space, X, d, then determine its underlying topological space $X, \mathcal{J}$, and finally ask for the comparison between "the locus..." in the *metric space* X, d and the open neighborhoods in the *topological space* $X, \mathcal{J}$. [Of course, this will just represent a rough check that we are on the right track. There are, as we have seen, topological spaces which do not arise from *any* metric space. Nonetheless we permit ourselves the definition "open neighborhood" in *every* topological space.] The comparison results in the following.

Theorem. Let X, d be a metric space, $X, \mathcal{J}$ its underlying topological space, p a point of X and ϵ a positive number. Let A be the set of all points q of X, with $d(p, q) < \epsilon$. Then, in the topological space $X, \mathcal{J}$ A is an open neighborhood of p.

Proof: Since $(p, p) = 0$, which is certainly less than ϵ, p is in A. So, we need only show that A is open in the topological space $X, \mathcal{J}$, i.e., that A is self-interior in the metric space X, d. Since, by an earlier theorem, int $(A) \subset A$, we have only to show that every point of A is in int (A). Let q be any point of A, and set $d(p, q) = a$, so $a < \epsilon$. Choose, by this last inequality, positive number b such that $a + b < \epsilon$. Now let r be any point of X with $d(q, r) \leq b$. Then, by the triangle inequality (since $d(p, q) = a$ and $d(q, r) \leq b$), we have $d(p, r) \leq a + b$. But $a + b < \epsilon$, and so $d(p, e) < \epsilon$, and so r is in A. So, we have shown that every point within distance b of point q is in A; hence, that q is in int (A). Since every point of A is in int (A), we have $A =$ int (A).

This is a slightly tricky proof. We are trying to show that A is an open neighborhood of p, and A is defined using the metric space. What we must show, then, is that p in A (which is easy), and that A is open (which is more difficult). But "open" in our topological space means "self-interior" in the metric space. It all reduces, therefore, to showing that every point of A is in int (A). For this, we select any point q of A, and proceed to show that this q must be in int (A), i.e., proceed to find a positive to show number b such that every point within b of q is in A.

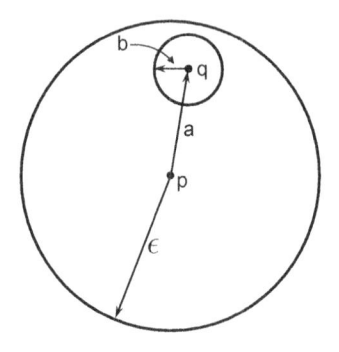

Note, however, that a "b" which works depends on what q is. One could have guessed that this would happen. Recall, way back, our demonstration that, for the disk in the plane, the interior of this disk is just the disk. There, we picked any point of the original disk, and found a smaller disk, centered at that point, which lied entirely within the original disk. Think of p as the center of the original disk, and of q as some point in that disk. How small must our "smaller disk" be, to ensure that it will be a subset of the original disk? Clearly, "how small" depends on what q is. The radius, b, of the smaller disk must be such that $a + b < \epsilon$. This observation about the plane, then, suggests the choice of b in the proof above.

The theorem above shows that "the locus ..." in a metric space always yields an open neighborhood in the underlying topological space. But what about the reverse? How do we pass from open neighborhoods in the underlying topological space to "the locus ..."? The other half of the comparison is the following.

Theorem. Let X, d be a metric space, $X, \mathcal{T}$ its underlying topological space, and p a point of X. Let A be an open neighborhood of p in the topological space $X, \mathcal{T}$. Then there exists a positive number ϵ such that every point q of X with $d(p, q) < \epsilon$ is in A.

Proof: Since A is an open neighborhood of p, p is a point of A. Further, A is open, i.e., A is self-interior in the metric space X, d, i.e., $A = \text{int}(A)$. So, since p is in A, p is in int (A). That is, there is a positive number ϵ such that every point q with $d(p, q) \leq \epsilon$ is in A. So, every q with $d(p, q) \leq \epsilon$ is certainly in A. This half is somewhat easier.

The situation, to summarize, is the following. We begin with metric space X, d and find its underlying topological space, $X, \mathcal{T}$. Fix point p of X. Then, for *any* positive ϵ, "the locus of points q of X with $d(p, q) < \epsilon$" (something defined using the metric space) is always an open neighborhood of p in the topological space. The reverse is that, given any open neighborhood of p in the topological space, there is *some* positive ϵ such that every point within ϵ of p is in that open neighborhood (but, of course, the neighborhood need not *be all* points q with $d(p, q) < \epsilon$ – it just has to *include* all such points). Using these two theorems, then, we can pass back and forth between "the locus ..." and "open neighborhood" – provided we are dealing with a metric space and its underlying topological space. [Of course, if we just dealing with a topological space – with no metric space around – then we do not want to "pass back and forth", since we do not have any metric space to "pass" to: We cannot even say "the locus ...".]

We turn now to our first definition – that of "interior" in a topological space. Recall, (page 53) the definition of "interior" for a metric space. What it says, essentially, is that point p gets to be in int (A) if, for some positive number ϵ, "the locus of points q with $d(p, q) < \epsilon$" is a subset of A. But our "generalized locuses", for a topological space, are the open neighborhoods. This immediately suggests how we should make our new definition.

Definition. Let $X, \mathcal{T}$ be a topological space, and $A \subset X$. Then the *interior* of A is the set of all points p of X such that some open neighborhood of p is a subset of A.

The first thing to note about this definition is that it makes sense. It does not refer to distances, i.e., to things we have only in metric spaces. Rather, it refers to the one thing we have access to in a topological space, namely, the open sets (or, what is practically the same thing, the open neighborhoods).

Recall what happened in the passage from the plane to metric spaces. We first defined the "interior" for a set in the plane. Then, once we got metric spaces, we defined "interior" for any subset of any metric space. One example of a metric space is of course the plane, with usual geometrical distance. Thus, we could ask ourselves the following question. Let A be a set in the plane. Then, using the definition on page 16, we can determine the interior of this set in the plane. But we could also do things in a somewhat different way. We might, alternatively, regard our plane, with geometrical distance, as a metric space. Then, we have a metric space X, d (with X the plane, and d geometrical distance), and we have $A \subset X$. Now, we can turn to the definition on page 53, and determine the interior of this subset of a metric space. The question is: Do we obtain the same "interiors", doing it in the two ways? The answer, of course, is that we do, as seen at the top of page 54. Now, we are passing from metric spaces to topological spaces. We can ask a similar question. Suppose we have a metric space X, d, and $A \subset X$. Then we can compute the interior of A, using the definition on page 53. However, we can alternatively do things in a different way. We can take the underlying topological space of our metric space. Then, we have $X, \mathcal{J}$, a topological space, and $A \subset X$. Now, using the definition on the previous page, we can determine the interior of A. Question: Do we obtain the same answer as we got from the definition on page 53. That is, do we obtain the same "interior" if we regard A as a subset of the metric space, and use the definition for a metric space, as if we regard A as a subset of the topological space, and use the definition for a topological space? If not, things would get pretty confusing, because we would always have to specify "which interior". [In fact, in practice, if we could obtain different answers by the two methods, one would probably use a different word for the definition on the previous page, say "interior".] Fortunately, things work out.

<u>Theorem</u>. Let X, d be a metric space, and $X, \mathcal{J}$ its underlying topological space, and $A \subset X$. Then the interior of A in the metric space X, d is the same set as the interior of A in the topological space $X, \mathcal{J}$.

Proof: Write $\text{int}_d (A)$ and $\text{int}_{\mathcal{J}} (A)$ for the two interiors. Let p be any point of $\text{int}_d (A)$, so, for some positive number ϵ, every point q with $d(p, q) \leq \epsilon$ is in A. That is, denoting by B the set of all q with $d(p, q) < \epsilon$, B is a subset of A. But, by the theorem on page 82, B is an open neighborhood of p. So, p is in $\text{int}_{\mathcal{J}} (A)$. For the converse, let p be any point of $\text{int}_{\mathcal{J}} (A)$, so there is an open neighborhood C of p, with $C \subset A$. By the theorem on page 83, there is a positive number ϵ such that every q with $d(p, q) < \epsilon$ is in C. So, every q with $d(p, q) \leq \epsilon/2$ is in C. But $C \subset A$, and so every q with $d(p, q) \leq \epsilon/2$ is in A. That is q is in $\text{int}_d (A)$.

Note how, in the proof, one just uses the two theorems which relate "open neighborhood" with "the locus of points ...". The "$\epsilon/2$ – business" near the end of the proof is necessary to switch from "<" (which comes from the theorem on page 83) to "≤" (which is needed in the definition on page 53). One does occasionally have to make such switches, and it is always done in this way (divide by 2).

What this theorem says, then, is that our definition of "interior" on page 83 is a reasonable one. The thing there defined deserves to be called "interior", for it is a genuine generalization of "interior" for metric spaces in the sense that, when both definitions are applicable (i.e., when we work with the underlying topological space of a metric space), they agree. For X, $\mathcal{T}$ *any* topological space, and $A \subset X$, we shall write int (A) for the interior of A. [The use of the same word, "interior", and the same symbol, "int ( )", for metric and topological spaces leads to no confusion, just as there was none in the passage from the plane to a general metric space.]

The theorem on the previous page produces instantly a long list of examples of interiors for sets in various topological spaces. Consider any of our earlier examples of metric spaces. Take the underlying topological space. Let $A \subset X$. Then int (A) (regarding A as a subset of this space) is just int (A) (regarding A as a subset of the metric space). For instance, let X be the plane, and $d(p,q) = 0$ if $p = q$ and 1 otherwise. Then, by the theorem on page 55, we have int $(A) = A$ for any subset A of this metric space. Now consider the underlying topological space. By the theorem on the previous page, we must have int $(A) = A$ (now, using the definition for "interior" on page 83) for any subset A of this topological space.

We give a few more examples.

Let X be any set, and X, $\mathcal{T}$ the discrete topological space. [That is *every* subset of X is deemed an open set in this topological space.] What are the open neighborhoods in this topological space? An open neighborhood of a point is an open set containing the point. Since every subset of X is open, an open neighborhood of p is any subset of X containing the point p. In particular, the subset of X consisting of p alone is an open neighborhood of p in this topological space. Now let $A \subset X$. We wish to determine the interior of this subset of this topological space. Suppose first that p is a point of A. Then, we claim, p is in int (A), that is, there is an open neighborhood of p which is a subset of A. Indeed, consider the open neighborhood of p consisting of p itself. This one is certainly a subset of A (since p is in A). So all the points of A are in int (A). What about points not in A? These can never be in int (A), and in fact this is true for *any* subset of *any* topological space.

Theorem. Let X, $\mathcal{T}$ be a topological space, and $A \subset X$. Then int $(A) \subset A$.

Proof: Let p be a point of int (A), so there exists an open neighborhood C of p with $C \subset A$. But p is in C and $C \subset A$. But p is in C and $C \subset A$, and so p is in A.

[Compare with the first theorem on page 67 Obviously, we are using virtually the same proof, essentially replacing "locus of points $q\ldots$" by "open neighborhood". It should be emphasized, however, that the theorem above is not an immediate consequence of that on page 67 One is for metric spaces, using the definition on page 53; the other for topological spaces, using the definition on page 83. Not all topological spaces come from metric spaces.] In any case, we conclude that, for A any subset of the topological space under consideration (that introduced on the previous page) int $(A) = A$. This, of course, is the answer we would have expected, from the theorem on page 84. Recall that for the metric space with distance "zero if the points are equal, one otherwise", the underlying topological space is discrete. But in such a metric space, int $(A) = A$ always; hence, also for the underlying topological space.

Let X be any set, and X, $\mathcal{J}$ the indiscrete topological space. [That is, the only subsets of X deemed open are the empty subset and X itself.] What are the open neighborhoods in this topological space? An open neighborhood of p is an open set containing p. But the only open sets are the empty set and X itself, while the empty set does *not* contain the point p. So, in this topological space, there is just one open neighborhood of a given point p, namely X itself. Now let $A \subset X$. We determine int (A). Given a point p of X, when will there be an open neighborhood of p which is a subset of A? But this is easy, since the only open neighborhood of p is X itself. If $A = X$, then clearly our open neighborhood of P will be a subset of A – for *every* point p of X. Thus, we have int $(X) = X$. Suppose, however, that A is not all of X. Then there *cannot* be an open neighborhood of p which is a subset of A (since the only open neighborhood of p is X itself, and this is not a subset of A). So, no point p will get to be in int (A). So, int (A) will be the empty set. Thus, for this topological space, int (A) is X if $A = X$, and the empty set otherwise. [Note that int (A) is always a subset of A, as required by the theorem on the previous page.]

We next establish some theorems for interiors. Many of these will be just the "topological versions" of earlier results for metric spaces. We emphasize again, however, that each theorem must now be proved anew, using the definition of a topological space, and of the interior of a subset of a topological space. But, of course, one uses the proofs for metric spaces as guides.

Theorem. Let X, $\mathcal{J}$ be a topological space, and $A \subset X$. Then int (A) is given by the union of all open subsets of A.

Proof: Point p is in int (A) if and only if p is a point of an open subset of A. [I hope the proof is clear. Point p being in int (A) just means that it is a point of an open subset of A. But, clearly, the set of points which are points of open subsets of A is just the set of points which are in the union of open subsets of A.] Note that "the union of all open subsets of A" is, by the third condition for a topological space, an open set, and this is of course a subset of A. Thus, what this theorem means is that int (A) is the largest open subset of A. Thus, to determine int (A), all one has to do is look around for open subsets of A, and find the largest one.

Theorem. Let X, $\mathcal{T}$ be a topological space, and $A \subset X$. Then A is open if and only if int $(A) = A$.

Proof: Let int $(A) = A$. Then, by the theorem above, int (A) is open, and so A is open. Let A be open. Then any point p of A has an open neighborhood which is a subset of A (namely, A itself), and so $A \subset$ int (A). But by the theorem on page 85, int $(A) \subset A$. So, int $(A) = 1$

[This is a bit strange. We motivated our definition of a topological space by looking for properties of self-interior sets in a metric space, and then demanding those properties for a collection of subsets of X in the definition of a topological space. Again using the analogy with metric spaces, we defined the interior for a subset of a topological space. So, we end up with topological spaces, and interiors therein. Nothing stops us, then, from looking at the "self-interior" sets in a topological space. When we do so, we find out that they turn out to be exactly the open sets in our topological space! Note that this theorem does *not* follow directly from any of the things about metric spaces. The definitions for topological spaces stand on their own, and things must be proven therein afresh.]

Theorem. Let X, $\mathcal{T}$ be a topological space. Then the interior of X is X, and the interior of the empty set is the empty set.

Proof: Since X and the empty set are open, by the first condition in the definition of a topological space, this is immediate from the preceding theorem.

Theorem. Let X, $\mathcal{T}$ be a topological space, and $A \subset X$. Then int (int (A)) = int (A).

Proof: By the theorem on page 86, int (A) is open. So, by the first theorem on page 87, int (int (A)) = int (A).

[Compare, the second theorem on page 67. For the corresponding theorem in metric spaces, we used the triangle inequality. It is as though the triangle inequality is somehow "hidden" in our conditions for a topological space. It seems remarkable that this should be possible, for the triangle inequality seems so "numerical", while the conditions for a topological space are so "set-theoretic". But the definition of a topological space clearly retains enough of the flavor of the triangle inequality to recover such results.]

Theorem. Let X, $\mathcal{T}$ be a topological space, and $A \subset B \subset X$. Then int $(A) \subset$ int (B).

Proof: Let p be a point of int (A), so there is an open neighborhood C of p with $C \subset A$. But, since $A \subset B$, this open neighborhood C of p also satisfies $C \subset B$. So, p is in int (B).

[Compare, first theorem on page 68.]

Theorem. Let X, $\mathcal{T}$ be a topological space, and $A \subset X$ and $B \subset X$. Then int $(A \cap B)$ = int $(A) \cap$ int (B).

Proof: Let p be a point of int$(A \cap B)$, so there is an open neighborhood C of p with $C \subset A \cap B$. But, since $C \subset A \cap B$, we have $C \subset A$ (and so p is in int (A)), and $C \subset B$ (and p is in int (B)). Hence, p is in int$(A) \cap$ int (B). For the converse, let p be a point of int $(A) \cap$ int (B). Then p is in int (A), and so there is an open neighborhood C_1 of p with $C_1 \subset A$; and p is in int (B), and so there is an open neighborhood C_2 of p with $C_2 \subset B$. Set $C = C_1 \cap C_2$, so certainly point p is in C, and $C \subset A \cap B$. But, by the second condition for a topological space, C is open. Thus, C is an open neighborhood of p, which $C \subset A \cap B$. So, p is in int $(A \cap B)$.

[Compare, the second theorem on page 68, whose proof is on page 30. Note that the proof of the analogous result for metric spaces requires that one take the smaller of two positive numbers, ϵ_1 and ϵ_2. In the topological case, these ϵ's get replaced by open neighborhoods, C_1 and C_2. Instead of taking the smaller of the ϵ's, we take the intersection of C's. The result will again be an open neighborhood, by the second condition for a topological space. Again, we have managed to replace a "numerical operation" (taking the smaller of two numbers) by a "set-theoretic operation" (taking the intersection of two sets). The definition of a topological space retains enough of the flavor of "comparison of distances" to permit us to recover this result.]

Theorem. Let X, $\mathcal{J}$ be a topological space, and $A \subset X$ and $B \subset X$. Then int $(A) \cup$ int $(B) \subset$ int $(A \cap B)$.

Proof: Since int $(A) \subset A$ and int $(B) \subset B$, int $(A) \cup$ int $(B) \subset A \cup B$. That is, int$(A) \cup$ int (B) is an subset of $A \cup B$. The result now follows from the theorem on page 86.

[Compare, theorem on page 68.]

We turn nest to the definition of "boundary" for a subset of a topological space. As usual, we use metric spaces as a guide. The definition of "boundary" for a subset of a metric space is on page 53. essentially, all we need do is replace "locus of points q with $d(p,q)\epsilon$" by "open neighborhood of p".

Definition Let X, $\mathcal{J}$ be a topological space, and $A \subset X$. Then the *boundary* of A is the set of all points p of X such that, for every open neighborhood C of p, there is a point of C in A, and also a point of C not in A.

Note that the definition even become a bit easier to state for topological spaces.

Immediately after defining "interior" for topological spaces, we carried out a check on the definition. We showed that, *for the case in which the topological space is the underlying one of a metric space*, the definition reduces to that of "interior" in a metric space. We can carry out a similar check for "boundary".

Theorem. Let X,d be a metric space, X, $\mathcal{J}$ its underlying topological space, and $A \subset X$. Then the interior of A in the metric space X,d is the same set as the boundary of A in the topological space X, $\mathcal{J}$.

Proof: Write bnd$_d$ (A) and bnd$_\mathcal{J}$ (A) for the two boundaries. Let p be a point of bnd$_d$ (A), so, for every positive number ϵ, there is a point q of A with $d(p,q) \leq \epsilon$, and also a point q' not in A with $d(p,q') \leq \epsilon$. Let C be any open neighborhood of p. Then, by the theorem on page 83, there is a positive number ϵ such that every point r with $d(p,r)\epsilon$ is in C. But, since p is in $\operatorname{bnd}_d(A)$, there is a point q in A with $d(p,q) \leq \epsilon/2$ (hence, with $d(p,q) < \epsilon$, hence, with q in C), and also a point q' not in A with $d(p,q') \leq \epsilon/2$ (hence, with $d(p,q') < \epsilon$, hence, with q' in C). So, for every open neighborhood C of p, there is a point of C in A, and also a point of C not in A. So, p is in $\operatorname{bnd}_{\mathcal{J}}(A)$. For the converse, let p be a point of $\operatorname{bnd}_{\mathcal{J}}(A)$. Let ϵ be any positive number, and let C be the set of points q with $d)p,q) < \epsilon$. Then, by the theorem on page 82, C is an open neighborhood of p. Since p is in $\operatorname{bnd}_{\mathcal{J}}(A)$, there is a point q of C (i.e., with $d(p,q) < \epsilon$, hence with $d(p,q) \leq \epsilon$) in A, and also a point q' of C (i.e., with $d(p,q') < \epsilon$, hence with $d(p.q') \leq \epsilon$) not in A. So, for every positive number ϵ, there is a point within ϵ of p and in A, and also a point within ϵ of p and not in A. So, p is in $\operatorname{bnd}_d(A)$.

It is just like the similar result for "interior". We again use our two theorems, on page 82 and 83, to convert from "the locus of point $q\ldots$" to "open neighborhood".

Again, we instantly obtain a long list of examples of boundaries of sets in topological spaces. We begin with any of our metric spaces, choose a subset A, and take the underlying topological space. Then the interior of A in this topological space is exactly the same set at the boundary of A in the original metric space. Again, we denote by $\operatorname{bnd}(A)$ the boundary of a subset of a topological space.

We give a few more examples.

Let X be a set, and $X, \mathcal{J}$ the discrete topological space (every subset of X open). Let $A \subset X$. In order that point p be in $\operatorname{bnd}(A)$, it must be the case that every open neighborhood of p contain a point of A, and also a point not in A. But one open neighborhood of p is the set consisting of p itself (certainly open in this topological space, and certainly containing the point p). But *this* open neighborhood C could hardly contain a point of A *and* a point not in A, for C only has one point in it. So, no point of X satisfies the condition for being in $\operatorname{bnd}(A)$. So, in this case, $\operatorname{bnd}(A)$ is the empty set. Again, compare with the theorem on page 55.

Next, let X be any set, and $X, \mathcal{J}$ the indiscrete topological space. Let $A \subset X$. In order that p be in $\operatorname{bnd}(A)$, it must be the case that every open neighborhood of p contain a point of A, and also a point not in A. But, in this topological space, the only open neighborhood of p is the entire set X. So, in order that p be in $\operatorname{bnd}(A)$, there must be a point of X in A, and also a point of X not in A. Suppose first that A is the empty set. Then there will never be a point of X in A. So, every point p will fail to satisfy the condition for being in $\operatorname{bnd}(A)$. So, $\operatorname{bnd}(A)$ will be the empty set. Next, suppose that A is all of X. Then there will never be a point of X not in A. So, every point p will fail to satisfy the condition for being in $\operatorname{bnd}(A)$. So, $\operatorname{bnd}(A)$ will be the empty set. Finally, suppose that A is neither the empty set nor X. Then there will always be a point of X in A (since A is not empty), and always a point of X not in A (since A is not all of X). So, *every* point p of X will satisfy the conditions for being in bnd (A). So, bnd (A) in this case will be X. Hence, for the indiscrete topological space, bnd (X) and the boundary of the empty set are empty, while the boundary of any other subset is X.

We give one final example of the definitions of "interior" and "boundary". Let X, $\mathcal{J}$ be the topological space introduced on pages 78 and 79 (in which X is the plane, and open sets are X, the empty set, and subsets of X consisting of all of X except for a finite number of points). Let p be a point of X. What are the open neighborhoods of p? The empty set, since it does not contain p, will never do. The set X is of course an open neighborhood of p. Furthermore, all of X except for a finite number of points (provided, of course, that the point p was not left out by being in that finite number!) will be an open neighborhood of p. So, we have the open neighborhoods. Now, let $A \subset X$. Let us try to find int (A). Point p will be in int (A) provided there is an open neighborhood of p. (i.e., all of X, or all of X except for a finite number of points) which is a subset of A. Suppose, for example, that A is our usual disk. Then no point p will satisfy this criterion (for A is just a little old disk: X itself is certainly is not a subset of A, and X except for a finite number of points is not a subset of A either). So, for this A, int (A) is empty. Clearly, int (A) will always be the empty set unless int (A) is "large enough" to contain one of these open neighborhoods. That is, int (A) will be empty unless A is "at least as large as all of X except for a finite number of points". Suppose, then, that set A is all of X except for a finite number of points. Then, of course, A is open in this topological space. But in this case, by the first theorem on page 87, int $(A) = A$. So, in this topological space, int (A) is the empty set unless A consists of all of X except for a finite number of points (for "smaller" A's than these do not contain any open neighborhoods as subsets), while, if A does consist of all of X except for a finite number of points, int $(A) = A$ (for A is then open). Next, we determine some boundaries. In order that p be in bnd (A), it must be the case that every open neighborhood of p contains a point in A and also a point not in A. The only open neighborhoods of p are X itself and all of X except for a finite number of points (where p is not one of that "finite number" left out). Let, for example, A be our disk. Then obviously every one of these open neighborhoods will contain a point in A and also a point not in A (for the open neighborhood must be "practically all of X"). So, for A the disk, bnd (A) will be X (for every point of X will satisfy the condition). Suppose, however, that A is somewhat "larger", say all of X except for a finite number of points, $p_1 \ldots, p_n$. Let p be a point not one of the p_i. Will it be true that every open neighborhood of p contains a point in A and also a point not in A? It will not be true: A itself is an open neighborhood of p, while A certainly does not contain a point not in A. So, the points of A will not be in bnd (A). However, let p be one of the p_i's, say $p = p_1$. Will it be true that every open neighborhood of p contains a point in A and also a point not in A? It will be true: Given any open neighborhood of p, it will certainly contain a point in A (since A is all of X except for a finite number of points, while any such open neighborhood must also be all of X except for a finite number of points). But such a neighborhood will also contain a point not in A, namely the point p itself. So, if p is one of the p_i, then p is in bnd (A). In this example, then, bnd (A) excludes all points in A, but includes the points $p_1 \ldots, p_n$. In other words, bnd $(A) = A^C$. Suppose, next, that A consists of just a finite collection of points, say $p_1 \ldots, p_n$. Let point p be none of the p_i. Is it true that, for every open neighborhood of p, it contains a point of A and also a point not in A? It is not true. Let C be all of X except for $p_1 \ldots, p_n$. This is an open neighborhood of p. But it contains no point in A (for we left out of C just the n points of A). So, this point: p is not in bnd (A). Let p be one of the p_i, say $p = p_1$. Now, every open neighborhood of p certainly contains a point in A (since the open neighborhoods are practically all of X, while "not in A" is also practically all of X), while every such open neighborhood also contains a point in A (namely p itself). So, this point p is in bnd (A). Thus, in this case, bnd $(A) = A$. To summarize, for this topological space, bnd (A) is A^C if A is all of X except for a finite number of points, bnd (A) is A if A is a finite subset of X, and bnd (A) is X otherwise.

We now obtain a few theorems involving boundaries.

Theorem. Let $X, \mathcal{J}$ be a topological space, and $A \subset X$. Then bnd (A) = bnd (A^C).

Proof: Point p is in bnd (A) if and only if every open neighborhood of p contains a point in A and also a point not in A. But points in A are just those not in A^C. So, point p is in bnd (A) if and only if every open neighborhood of p contains a point not in A^C and also a point in A^C, i.e., if and only if p is in bnd (A^C).

[Compare, the theorem on page 68, proven on page 29. It is almost exactly the same.]

Theorem. Let $X, \mathcal{J}$ be a topological space, and $A \subset X$. Then every point of X is in int (A) or bnd (A) or int (A^C), and no point of X is in more than one of these sets.

Proof; Let p be a point of X. If some open neighborhood of p is a subset of A, then p is in int (A). If some open neighborhood of p contains no points of A, then p is in int (A^C). Clearly, p cannot be in both int (A) and int (A^C). If neither of these holds, i.e., if every open neighborhood of p contains a point not in A and also a point in A, then p is in bnd (A). Finally, if p is in bnd (A), then (since every open neighborhood of p contains a point not in A) p cannot be in int (A), and (since every open neighborhood of p contains a point not in A^C) p cannot be in int (A^C).

[Compare, the theorem on page 68.]

These two results, at least for the plane, seemed "geometrically reasonable", in that they fit in with our intuitive ideas of the words "boundary" and "interior". But they continue to hold even in topological spaces! Again, the definition of a topological space has somehow managed to capture our intuitive ideas.

As an example of these two theorems, consider the topological space discussed on page 91 ("all of X except for a finite number of points"). Let $A \subset X$ be a finite set. Then bnd $(A) = A$, as we saw. Now, A^C is all of X except for this finite set of points. So, as we saw above, bnd A^C) is that finite set, i.e., bnd $(A^C) = A$. That is, we have bnd $(A) =$ bnd (A^C), as demanded by the theorem above. In this example, int (A) is the empty set, and int (A^C) is A^C. So, int $(A) =$ empty, bnd $(A) = A$, and int $(A^C) = A^C$. It is indeed true, as guaranteed by our theorem on the previous page, that every point is in one of these three sets, and no point is in more than one. Alternatively, for A the disk, int$(A) =$ empty, bnd $(A) = X$, and int $(A^C) =$ empty. Again, every point of X is in one and only one of these sets.

Finally, we turn to "connected" for topological spaces. We can just take over directly the definition for metric spaces, page 60

Definition. Let X, $\mathcal{T}$ be a topological space, and $A \subset X$. Then A is said to be *disconnected* if there exists a subset B of X such that some point of A is in B, some point of A is not in B, and no point of A is in bnd (B).

Of course, "bnd (B)" in this definition refers to the boundary *for a subset of a topological space*. [It could hardly be anything else. We do not have any "distances" now, to use for the definition of boundary for a metric space.]

Again, we check that our definition agrees with that for metric spaces when both are applicable.

Theorem. Let X, d, be a metric space, X, $\mathcal{T}$ its underlying topological space, and $A \subset X$. Then A is disconnected in the metric space X, d if and only if A is disconnected in the topological space X, $\mathcal{T}$.

Proof: A is disconnected in the metric space if and only if there exists a subset B of X such that some point of A is in B, some point of A is not in B, and no point of A is in bnd$_d$ (B). But, by the theorem on page 88, bnd$_d$ $(B) =$ bnd$_\mathcal{T}$ (B). So, A is disconnected in the metric space if and only if there exists a subset B of X such that some point of A is in B, some point of A is not in B, and no point of A is in bnd$_\mathcal{T}$ (B), i.e., if and only if A is disconnected in the topological space.

Again, the theorem produces immediately examples, from those for metric spaces, of connected and disconnected subsets of various topological spaces. We do our usual two other examples.

Let X, $\mathcal{T}$ be a discrete topological space. Then for any $A \subset X$, bnd (A) is the empty set. Let $A \subset X$. When will A be disconnected? We must find a set B such that some point of A is in B, some point of A is not in B, and no point of A is in bnd (B). But the last will always hold, no matter what B is, for bnd (B) is the empty set. If A is the empty set, or A consists of just one point, then A will be connected. [For, under these circumstances, one could hardly find B such that some point of A is in B *and* some point of A is not in B.] Suppose, however, that A consists of two or more points. Let B be the set consisting of just one point, say p, of A. Then some point of A is in B (namely, p), some point of A is not in B (since A contains two or more points, while B consists of just p), and no point of A is in bnd (B) (since bnd (B) is empty). So, such an A is disconnected. Thus, in a discrete topological space, the empty subset and a subset consisting of just one point are both connected, while any other subset is disconnected.

Let X, $\mathcal{T}$ be an indiscrete topological space. Then bnd (B), for $B \subset X$, is the empty set if B is empty or $B = X$, and bnd $(B) = X$ otherwise. Now let $A \subset X$. When will A be disconnected? When will one be able to find B such that some point of A is in B, some point of A is not in B, and no point of A is in bnd (B)? We claim that one will never find such a B (no matter what A is). Indeed, since some point of A is supposed to be in B, B cannot be the empty set; since some point of A is supposed to not be in B, B cannot be X itself. But if B is neither of these, then bnd $(B) = X$. How can it happen, then, that no point of A is in bnd (B)? Only if A is the empty set. But finally, if A is the empty set, then it could hardly be that some point of A is in B. Thus, no matter what A is, we shall never be able to find a suitable set B in this topological space. In short, *every* subset of this topological space is connected. [This includes even, for example, an A consisting of just two points of X. This set, in this topological space, is connected!]

Finally, we recover the old theorem about connected sets.

Theorem. Let X, $\mathcal{T}$ be a topological space, A and B connected subsets of X, and p a point of both A and B. Then $A \cup B$ is connected.

Proof: Suppose, for contradiction, that some point of $A \cup B$ (say, u in A) is in C, some point of $A \cup B$ (say, v in B) is not in C, and no point of $A \cup B$ is in bnd (C), for some $C \subset X$. Then no point of A and no point of B is in bnd (C). But point u in A is in C and no point of A is in bnd (C), and so, since A is connected, every point of A must be in C. In particular, point p must be in C. But now point p in B is in C, point v in B is not in C, and no point of B is in bnd (C), which contradicts the fact that B is connected.

Again, the proof is just the old proof (page 33). Again, an "intuitively reasonable" theorem is recovered within the context of topological spaces.

Finally, we remark that all our old examples for the plane, and for metric spaces, become also examples for topological spaces, since the plane is a metric space and metric spaces give rise to topological spaces. In particular, all the things we found to be false before (provided those "things" use only interior, boundary, etc.) remain false for topological spaces.

I hope that some sense of the beauty of the definition of a topological space comes through all this. We put so little in (just give a set X and some subsets), and yet so much seems to come out. We shall see more of this shortly.

11. An Example: Connected Sets with Boundary Attached

All of the theorems we have dealt with so far have been relatively simple ones. This was necessary in order that complications inherent in the theorems themselves not obscure the transitions from the plane to metric spaces to topological spaces. We here give one example of a more complicated theorem. It will also serve to illustrate the idea that the notion of a topological space indeed succeeds in capturing a wide range of intuitive ideas.

Let A be a set in the plane. Then, intuitively, the boundary of A is "right up against A"; it does not contain "parts which are well-separated from A". These intuitive observations suggest the following. Suppose that A is connected. Let us now "attach to A its boundary", i.e., consider the set $A \cup$ bnd (A): Since bnd (A) would seem to be "right up against A", one might expect that $A \cup$ bnd (A) will also be connected. Each point of bnd (A) is in such "intimate contact" with A that it is hard to see how the attachment of such points to A could make a connected set A become disconnected.

All this suggests

Theorem. Let X, $\mathcal{T}$ be a topological space, and A a connected subset of X. Then $A \cup$ bnd (A) is connected.

Our next job, as topologists, would be to try to think of a proof. One might try first the idea of showing that bnd (A) must be connected and that A and bnd (A) have a point in common. Then, one could use the theorem that the union of two connected sets having a point in common is connected. Unfortunately, this will not work. Consider, for example, A the usual disk in the plane. Then bnd (A) (the "rim" of the disk) will *not* have points in common with A. Furthermore, we have seen that it is not true that the boundary of a connected set is connected. To come up with a proof that works, one might go back and look at the intuitive ideas which suggested the theorem in the first place. Suppose for a moment that, with A connected $A \cup$ bnd (A) could be "separated into two separate pieces". Why might one expect this to be impossible? Since A is to be connected, if such a separation were possible, then all of A would presumably have to be in "one of the pieces" –

for otherwise we would have separated A into two pieces. Thus, one of our "pieces" would have to contain all of A (and possibly also some boundary points), while the other would have to contain *only* boundary points. But now, since these boundary points are not "well separated from A" (boundary points being as they are), it is hard to see how our two pieces could be "well separated", as required in the definition of disconnected. These ideas, written out more formally, generate the proof.

Proof: Suppose, for contradiction, that $A \cup$ bnd (A) were disconnected, so there is a subset C of X such that some point, p, of $A \cup$ bnd (A) is in C, some point, q, of $A \cup$ bnd (A) is not in C, and no point of $A \cup$ bnd (A) is in bnd (C). If p is in A, set $p' = p$. If p is not in A, then, since p is in $A \cup$ bnd (A), p must be in bnd (A). But p is in C and not in bnd (C), and so p must be in int (C). But int (C) is open, and so is an open neighborhood of p. Since p is in bnd (A), and int (C) is an open neighborhood of p, there exists a point, designated p', in A and int (C). But int $(C) \subset C$, and so p' is in A and C. Thus, in either case (whether p is in A or not), we obtain a point p' in A and in C. Similarly for q. If q is in A, set $q' = q$. If q is not in A, then q is in bnd (A). But q is in neither C nor bnd (C), and so must be in int (C^C). So, int (C^C) is an open neighborhood of q, and q is in bnd (A), and so there is a point, designated q', in int (C^C) and A. But int $(C^C) \subset C^C$, and so q' is in A and not C. Thus, in either case, we obtain a point q' in A and not C.

We now have a set C such that some point (namely, p') of A is in C, some point (namely q') of A is not in C, and no point of A is in bnd (C) (since no point of $A \cup$ bnd (A) is in bnd (C)). But this contradicts the fact that A is connected.

This theorem is perhaps a bit more typical than the others of a nontrivial theorem in topology. What is perhaps surprising about it is that, while there is so little in the definition of a topological space, such results are actually true.

12. Compactness

We have generalized "interior", "boundary", and "connected" from metric spaces to topological spaces. We now turn to the generalization of "bounded".

One sees immediately that "bounded" is going to be more difficult. Consider, for example the definition of "interior" for a metric space. It reads "The interior of A is the set of all points p such that, for some positive number ϵ...", where "..." refers only to the points within distance ϵ of p. So, the analogous definition for topological spaces becomes "The interior of A is the set of all points p such that, for some open neighborhood of p,...". That is, we may just mechanically replace "the locus of points within ϵ of p" by "open neighborhood of p". But consider, by contrast, the definition of a bounded subset of a metric space: "Set A is said to be bounded if, for every positive ϵ, there exists a finite set of points of X, $p_1, \ldots, p_n$, such that every point of A is within ϵ of at least one of these points." The critical difference here is that one has to know the value of ϵ *before* one knows the finite set of points, $p_1, \ldots, p_n$. Indeed, as we have seen, one has to select one's finite set of points carefully, depending on what ϵ is. What we are demanding for bounded is that every point of A be in the locus of points within ϵ of p_1, or in the locus of points within ϵ of p_2, or in the locus ... It is these locuses which we would like to replace by open neighborhoods. But how is this to be done, since we are to know the ϵ (i.e., the "sizes" of these open neighborhoods) prior to knowing which points they are to be open neighborhoods of? Thus, suppose one tries to write out an analogous definition of "bounded" for topological spaces: "Set A is said to be bounded if for every open neighborhood, there exists a finite set of points of X, $p_1, \ldots, p_n$, such that every point of A is in ...". How does one finish the sentence? Every point of A is in what? Our open neighborhood is just sitting there: We do not have open neighborhoods of $p_1, \ldots, p_n$. [If we did, then of course, one would demand that every point of A be in at least one of these open neighborhoods.] Furthermore, we say "if for every open neighborhood ...". Open neighborhood of what point? [Surely not of the p_i's, for we do not have access to the p_i's at this point in the definition.]

The key to finding something analogous to "bounded", but for topological spaces, is, as it turns out, to first reformulate somewhat the definition for metric spaces. This reformulation will be somewhat smoother using the following, easy, definition.

Definition. A collection of sets is said to *cover* set A if A is a subset of their union.

In other words, in order that a collection of sets cover A, it is necessary that *every* point of A be in at least one of those sets. [That is, it is essentially the same meaning as in "cover the wall with paint".]

Let X, d be a metric space, and $A \subset X$. It is easy to insert the word "cover" into the definition of boundedness of A. "Set A is said to be bounded if, for every positive ϵ, there exists a finite set of points of X, $p_1, \ldots, p_n$, such that the n sets, B_1 (the set of all points within ϵ of p_1), B_2, (the set of all points within ϵ of p_2), ..., B_n (the set of all points within ϵ of p_n) cover A." [This just says the same thing as before, for demanding that the B's cover A is the same as demanding that every point of A is within ϵ of at least one of the p_i's.]

Of course, the mere insertion of the word "cover" does not immediately solve our problem. The next step is the one that contains the idea. Let X, d be a metric space, and $A \subset X$. Let a positive number ϵ be given. For p any point of X, let us write B_p for the set of all points q of X with $d(p, q) \leq \epsilon$. In the plane, for example, B_p would be a disk, of radius ϵ, centered at the point p. Now let us consider for a moment *all* subsets of X of the form B_p for p *any* point of X. In the plane, for example, this would be an enormous collection of disks, of radius ϵ, but with one for each point of the plane, centered at that point. Here, then, is a collection of subsets of X. This collection of sets certainly covers A. [Indeed, it even covers X. Given *any* point p of X, it is in at least one of the B's, namely, in B_p. Of course, this p may also be in other B's, but one is enough in order to cover.] So, we so far have $A \subset X$, we have positive ϵ given, and we have found a collection of sets, the B_p's, which cover A. Nothing has been said yet about "bounded". We now ask: How can one express, in terms of the B_p's, the statement that A is bounded? But this is easy. The statement is that a *finite* number of the B_p's will cover A.

[Indeed, "The finite number of B's, $B_{p_1}, B_{p_2}, \ldots, B_{p_n}$, cover A." means the same thing as "Every point of A is within distance ϵ of at least one of the points $p_1, p_2, \ldots, p_n$." Thus, "*Some* finite number of B's cover A." is the same as "*There exists* a finite set of points, $p_1, \ldots, p_n$, such that every point of A is within ϵ of at least one of them."]

So far, then, we have the following. Let X, be a metric space, and $A \subset X$. Then "A is bounded" means the same thing as "For any positive ϵ, a finite number of the B_p's (constructed as above *using the given ϵ*) cover A." Now for the final step in the translation. The B_p's are, of course, unsuitable for topology, for they are defined using the distance, while we shall not have distances in a topological space. By what should they be replaced? But they are practically self-interior. [They would have been self-interior, had we defined them as "all points q with $d(p,q) < \epsilon$" instead of "$\leq \epsilon$". But this difference, as we have seen, is not significant in these matters, for one can always replace ϵ by $1/2\epsilon$.] Thus, the plan is to replace "the B_p's" (for metric spaces) by "open sets" (for topological spaces). In carrying out the replacement, however, we would wish to retain, in the topological case, as much of the flavor of the B_p's in the metric case as we can. But there seems to be only one property of the B_p's that we say without using distance, namely that the B_p's cover A. So, we decide to replace "the B_p's" by a "collection of open sets which cover A". By what shall we replace the part about "For any positive ϵ"? Of course, the only role of ϵ is to tell us what B_p's we are to choose. thus, "for any positive ϵ" could be replaced by "for any collection of B_p's". [We emphasize that none of these replacements and translations are at all obvious. In practice, one would of course have to try a vast number of things, to see what works and what does not. One could easily obtain three or four generalizations – or none.]

In any case, these remarks suggest the following.

Definition. Let X, $\mathcal{T}$ be a topological space, and $A \subset X$. Then A is said to be *compact* if, for any collection of open sets which covers A, a finite number of these sets cover A.

It is conventional, in topology, to use the word "compact" rather than "bounded" – indeed, as we shall see, the two words have a somewhat different meaning. Thus, to show that a subset of a topological space is compact, one has to show that, no matter what collection of open sets one is given which covers A, some finite number of those will suffice to cover A. To show A not compact, one must find a collection of open sets which covers A, such that no finite number will do.

Some examples and elementary results will illustrate the definition.

Theorem. Let X, d be a metric space, $X, \mathcal{T}$ its underlying topological space, and A a compact subset of X in this topological space. Then A is bounded in the metric space.

Proof: Let a positive ϵ be given. for p any point of X, denote by B_p the set of all points q with $d(p,q) < \epsilon$. Then, by the theorem on page 82, each B_p is open in X, $\mathcal{T}$. But the B_p's cover A, and so, by compactness, a finite number $B_{p_1}, \ldots, B_{p_n}$, cover A. That is, every point of A is within ϵ of at least one of the finite number of points, $p_1, \ldots, p_n$. Thus, compact sets in the underlying topological space are always bounded in the metric space. Compare, for instance, with the theorem on page 84. (You get the same answer whether you take the interior in the metric space or in the underlying topological space.) This might suggest that the converse of this theorem should also be true: That any bounded set in the metric space must be compact in the underlying topological space.

Unfortunately, this converse is not true. Let X, d be the usual plane, and A the disk consisting of all (x, y) with $x^2 + y^2 < 1$. Then, as we have seen, A is bounded in the metric space. Is A compact in the underlying topological space? Is it true that, given any collection of open sets which covers A, a finite number will do? For a any positive number less than 1, denote by C_a the set of points (x, y) with $x^2 + y^2 < a^2$, so C_a is "the disk of radius a centered at the origin". All of these C_a's are of course open in the underlying topological space. Clearly, they cover A. [Indeed, give any point p of A, so $d(p, \text{origin}) < 1$, choose a such that $d(p, \text{origin}) < a < 1$. Then p is in C_a.]

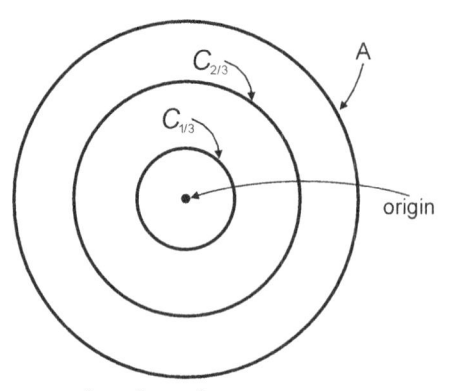

So, if A is to be compact, some finite number of these open sets had better cover A. But none will do. Indeed, suppose it was claimed that the finite number $C_{a_1}, \ldots, C_{a_n}$ cover A. Each of $a_1, \ldots, a_n$ must, of course, be a positive number less than one. Let a be the largest of these numbers, so a is less than one. Now choose point q with $a < d(q, \text{origin}) < 1$. Then this q will be in A, but will not be in any of our finite collection of C's. Thus, this disk is not compact.

Thus, at least for the plane, "compact" is stronger than just "bounded". However, it turns out that the disk is not compact for what is really a rather minor reason: We did not "attach to A its boundary". For sets "with boundary attached", in the plane, compactness means the same thing as boundedness. A simple, precise way to say "with boundary attached" is to demand that A^C be self-interior. Indeed, if A^C is self-interior, so $A^C = \text{int}(A^C)$, then since, by the second theorem on page 91, no point is in both bnd (A) and int (A^C), no point can be in both bnd (A) and A^C. That is, every point of bnd (A) must be in A. That is, A must include its boundary. We have

Theorem. Let X, d be the usual plane, and A a bounded subset of X with A^C self-interior. Then, in the underlying topological space, A is compact.

Unfortunately, a full proof of this requires some facts about real numbers which, while not terribly difficult, would require a rather lengthy explanation. So, we merely sketch the idea. Suppose, for contradiction, that A were not compact, so we have some collection of self-interior sets, B's, which cover A, but such that no finite number cover A. Cover A by a "grid" consisting of non-overlapping squares, each one unit by one unit. Since A is bounded, only a finite number of such squares will be needed.

We now ask, for each such square S: Does a finite number of the B's cover $A \cap S$? The answer could be yes or no. However, the answer could hardly be "yes" for *every* one of our squares, for, if a finite number of B's covered the "part of A in S" for *every* square S, then, since there are only a finite number of squares, a finite number of the B's would cover A. So, there are must be at least one square S such that no finite number of B's cover $A \cap S$, say, the square indicated.

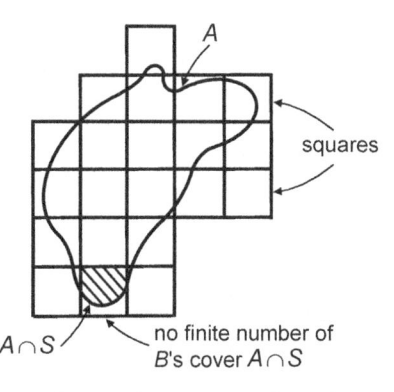

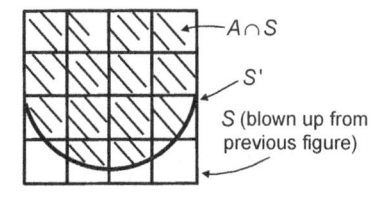

Now consider just this one square, and subdivide it into smaller squares, as shown. We now ask, for each of these smaller squares, S': Does a finite number of B's cover $A \cap S'$? Again, the answer cannot be "yes" for *every one* of the smaller squares, for if this were the case then a finite number of B's would have to cover $A \cap S$. So, choose one of the smaller squares, S', such that no finite number of B's covers $A \cap S'$. Now subdivide this S', into still smaller squares, ask the same question for each of these, and find the still smaller square for each the answer is "no". Continuing in this way, we find a succession of smaller and smaller squares, each inside its predecessor. Let p be the point which is common to all of these squares. [It is for the existence of such a point of the plane that we need a property of the numbers.] We now ask the crucial question: Where is p, in A or not in A? Either answer will get us into trouble. Suppose p were not in A, so p is in A^C. But A^C is self-interior, so p is in int (A^C). That is, there is a positive number ϵ such that no point within ϵ of p is in A. But now one of our "succession of smaller and smaller squares, converging down on p" must lie entirely within this locus of points within ϵ of p. Call this square C, so C is a subset of A^C, i.e., C has no points in common with A. But recall how this C was to have been chosen (from a subdivision of a larger square): It was to be the one such that no finite number of the B's covers $A \cap C$. But this is a contradiction, for $A \cap C$ is the empty set, and certainly a finite number of the B's (such as, for example, any one of the B's) covers the empty set. Thus, "p not in A" leads to a contradiction. Suppose, then, that p were in A. But the B's are supposed to cover A, so one of the B's, say B_1, must contain the point p. Each B is self-interior. So, there is a positive number ϵ such that every point within ϵ of p is in B_1. But again one of our "succession of smaller and smaller squares, converging down to p" must lie entirely within this locus of points within ϵ of p. Call this square C, so C is a subset of B_1. But again C was to have been chosen, from a subdivision of a larger square, as the one such that no finite number of B's covers $A \cap C$. This is again a contradiction, for some finite number of the B's *does* cover $A \cap C$ in this case, namely the finite number consisting of only B_1. (For C is a subset of B_1, so B_1 covers C, so certainly B_1 covers $A \cap C$.) So, "p in A" also leads to a contradiction. Thus, we began by assuming that we had a collection of self-interior sets which cover A, but such that no finite number would do, and we obtain a contradiction. Hence, for any collection of self-interior sets which cover A, some finite number must cover A. That is, A must be compact.

Thus, for the plane, "compact" is not quite the same thing as "bounded", but it is quite close. One only has to deal with sets "having their boundaries attached", and the two notions agree. We give some more examples.

Theorem. Let X, $\mathcal{J}$ be a topological space, and A a finite subset of X. Then A is compact.

Proof: Let the points of A be $p_1, \ldots, p_n$. Consider a collection of open sets which covers A. Then one, say B_1, must contain p_1; and one, say B_2, must contain p_2; and so on to B_n, containing p_n. But now $B_1, B_2, \ldots, B_n$ is a finite number of sets in our collection which covers A.

Recall that every finite subset of a metric space is bounded. Again, compact is "similar" to bounded.

Let X, $\mathcal{J}$ be indiscrete, and $A \subset X$. Then A is always compact. Suppose first that A is empty. Then *any* set covers A. So, given any collection of open sets which covers A, a finite number (say, any one) will do. Suppose that A is not empty. Consider a collection of open sets which covers A. Since X, $\mathcal{J}$ is indiscrete, the only open sets are X and the empty set. But, since A is to be covered by these sets, at least one of them had better be X (for a lot of empty sets won't cover a not-empty A). But now a finite number of our open sets will do, namely just one of them, the "X".

Let X, $\mathcal{J}$ be discrete, and $A \subset X$. If A is finite, then, by the theorem above, A is compact. Suppose, then, that A is not finite. Then, we claim, A is not compact either. Indeed for each point of A, consider the subset of X consisting of just that point. These sets certainly cover A, i.e., every point of A is in one of them. Since our topological space is discrete, every subset of X is open, so these sets are open. But no finite number of these sets could cover A, for each set only contains one point, so a finite number could only cover a finite set, while A is not finite.

The union of two bounded sets in a metric space is bounded. Similarly, Theorem. Let X, $\mathcal{J}$ be a topological space, and A and B compact subsets of X. Then $A \cup B$ is compact.

Proof: Let there be given a collection of open sets which covers $A \cup B$. Then this collection covers A, and so, since A is compact, some finite number,

$C_1 \ldots, C_n$, covers A. Also, this collection covers B, and so, since B is compact, some finite number, $C_1' \ldots, C_m'$, covers B. But now the finite number $C_1 \cdots, C_n, C_1' \ldots, C_m'$, covers $A \cup B$.

In a metric space, any subset of a bounded set is bounded. This is false for compact sets in a topological space. For example, let B be the set of all (x, y) in the plane with $x^2 + y^2 \leq 1$, the "disk with boundary attached". Then, by the theorem on page 100, B is compact in the underlying topological space. Let $A \subset B$ consists of all (x, y) with $x^2 + y^2 < 1$. Then A is a subset of a compact set, but A is, as we have seen, not compact. Again, the problem has to do with "attachment of boundaries". What is true, however, is that subset A of compact B, provided A "has its boundary attached" is always compact.

Theorem. Let X, $\mathcal{T}$ be a topological space, B a compact subset of X, and $A \subset B$ with A^C open. Then A is compact.

Proof: Let there be given a collection of open sets which covers A. Include in this collection the open set A^C. Then, since the original collection covers A, and A^C covers A^C, the resulting collection covers X. In particular, it covers B. Since B is compact, a finite number of sets from this collection covers B, say $C_1, \ldots, C_n$ and A^C, where the C's are from the original collection. Since $A \subset B$, it must be that $C_1, \ldots, C_n, A^C$ covers A. But, since no point of A is in A^C, if $C_1, \ldots, C_n, A^C$ covers A then $C_1, \ldots, C_n$ must cover A.

Thus, a finite number of sets from the original collection covers A.

This is a tricky proof. One begins with a collection, the C's, of open sets which covers A. One tosses in an extra open set, A^C. The resulting collection covers B, and so, by compactness of B, a finite number will do. Of course, in this "finite collection" there may be included the extra set we threw in, namely A^C. Since $A \subset B$, this finite collection also covers A. But A^C can now be left out again, for it is not going to help any in covering A, since it has no points in common with A. Thus, we find a finite number of C's to cover A.

To summarize, "bounded" undergoes a curious transition from metric spaces to topological spaces. As the definition of "bounded" stands in metric spaces, it looks as though it will be impossible to generalize. One does obtain a generalization of sorts, to compactness, but it is not quite the same thing. *Very* roughly speaking, compact sets are like bounded sets which include their boundaries. But, compactness at least has much of the flavor of boundedness. In particular, whenever, in an argument, one would like to "think boundedness", one instead "talks compactness". There are enough properties in common so that one usually gets by. What is nice about compactness (aside, of course, from the most remarkable thing: that there should even *exist anything* analogous to "bounded" in topological spaces) is that the definition is so simple, and the proofs are so simple. Indeed, the proofs of the previous two theorems are perhaps simpler than the corresponding proofs for boundedness in metric spaces. It turns out that compactness is an extremely important and useful idea in topology – much more so than, for example, connectedness. Again, an intuitive idea which could easily have been lost in the transition to topology in fact survives.

13. Continuous Mappings of Topological Spaces

One of the central ideas in topology is that of a continuous mapping. In this section, we define continuous mappings, give some examples, and obtain some elementary properties.

Recall, from Sect, 5, our treatment of continuous curves in the plane. A curve was a rule which assigned, to each number t a point $\gamma(t)$ of the plane. We then considered continuous curves, those which "do not skip". A curve was said to be continuous, roughly speaking, if nearby t-values were assigned, according to the rule γ, to nearby points of the plane. The notion of a continuous mapping of topological spaces is a direct generalization of these ideas. What is it that goes into a continuous curve? Well, there is the set of real numbers, the set of points of the plane, and the rule γ which assigns, to each point of the first set, a point of the second. The part about "continuous", however, uses more structure; "For every positive number ϵ there exists a positive number δ such that, whenever $|t - t_0| \leq \delta$, $d(\gamma(t), \gamma(t_0)) \leq \epsilon$." What is this $|t - t_0|$? It is essentially the "distance", in the set of numbers between the point t of that set and the point t_0. Indeed, as we have seen, the set of numbers with this distance forms a metric space. Similarly, $d(\gamma(t), \gamma(t_0))$ makes use of the distance between two points of the plane. In short, what really goes into the notion of a continuous curve is i) the *metric space* of real numbers (with distance $|t - t_0|$), ii) the *metric space* the plane (with usual geometrical distance), and iii) a rule γ which assigns, to each point of the first metric space, a point of the second metric space. But the generalization, in topology, of a metric space is a topological space. Thus, the ingredients of what we are looking for should be i) a *topological space*, ii) a second *topological space*, and iii) a rule which assigns, to each point of the first topological space, a point of the second. Just as, for a *continuous* curve, one requires a certain property of the rule, so, for a continuous mapping of topological spaces, one will require a certain property.

These introductory remarks out of the way, we proceed to the definitions. We first wish to generalize "curve". This, from the observations above, is straightforward. Let X and X' be sets. Then a *mapping* from set X to set X' is a rule γ which assigns, to each point of the first set, X, a point of the second set, X'. For p a point of X, we write $\gamma(p)$ for the point of X' assigned by this rule to p. It is also convenient to use the symbol of $X \xrightarrow{\gamma} X'$ to mean "γ is a mapping from set X to set X'". We emphasize that, to specify a mapping, one must i) say what the set X is, ii) say what the set X' is and iii) give the rule γ. This rule must of course be unambiguous, it just specify a point of X' for *any* point of X, and it must yield *just one* point of X' in exchange for a point of X. [In fact, we shall be using mappings primarily when X and X' are the sets for topological spaces. But we do not need the open sets to say what a mapping is (any more than we needed the distances to say what a curve is), and so there is no harm in defining "mapping" for any sets X and X'.] It is ok if a mapping have the feature that some points of X' are not assigned from *any* point of X, or if two different points of X are sent, by the mapping, to the *same* point of X'. As long as γ specifies unambiguously a point of X' in exchange for a point of X, and as long as this specification is given for any point of X, this γ is a mapping.

"Mapping" is perhaps the second most important idea in all of mathematics. [It would be difficult to unseat "set" from first place.] We give some examples of mappings.

A curve (page 36) is just a mapping from the set of real numbers (X) to the set of points of the plane (X').

What is usually called a "function of one variable" in elementary mathematics is a rule which assigns to each real number, x, another real number, written $f(x)$. That is, this is just a mapping from X (the set of real numbers) to X' (also the set of real numbers).

Let X be any set, and let $X \xrightarrow{\gamma} X$ be the rule which assigns, to point p of X, the point p of X. [That is $\gamma(p) = p$.] This rule unambiguously gives a point of the second set (X) in exchange for a point of the first set (also X), namely "keep the same point". This is a mapping.

Let X be any set, and X' the plane. Let $X \xrightarrow{\gamma} X'$ be the rule which assigns, to any point p of X, the origin of X' (the plane), i.e., $\gamma(p)$ = origin. This is a mapping. [Here, of course, all points of X get sent by the rule to the same point of X', and so most points of X' (namely, all but the origin) do not come from any point of X.]

Let X, d be any metric space, and fix once and for all a point p of X. Let X' be the set of numbers. Let $X \xrightarrow{\gamma} X'$ be the following rule: Given point q of X, set $\gamma(q) = d)p, q)$. This indeed assigns a point of X' (i.e., a number) to each point, q, of X. So, we have a mapping. Here, for example, $\gamma(p) = 0$, by the first condition for a metric space.

Let $Y, \mathcal{T}$ be any topological space. Let X be the set of all subsets of Y. [Thus, a "point" of the set X is a subset of Y.] Let X' be a set consisting of just two elements, which we may denote "yes" and "no". Now let $X \xrightarrow{\gamma} X'$ be the following rule: Given a point p of X (that is, given a subset of Y), let $\gamma(p)$ be the point "yes" of X' if this subset of Y is open, and "no" if it is not open. Here is a rule which assigns to each point of X (i.e., subset of Y), a point of X' (i.e., the point "yes" or "no"). This is a mapping.

Let $Y, \mathcal{T}$ be any topological space. Let X be the set of all subsets of Y. Let $X \xrightarrow{\gamma} X$ be the rule which assigns, to any point p of X, the point bnd (p) of X. [In more detail, "point" p is a subset of topological space $Y, \mathcal{T}$. We can take the boundary, of this subset, and again obtain a subset of Y. That is, we again obtain a "point" of the set X. This point is $\gamma(p)$.] This is a mapping.

Let X be the plane, and X' be the real numbers. Let $X \xrightarrow{\gamma} X'$ assign, to point (x, y) of X, the number $x^3 - 3xy$. This is a mapping. We shall later need two easy definitions with mappings.

Definition. Let X and X' be sets, and $X \xrightarrow{\gamma} X'$. Let $A \subset X$. Then the *image* of A under γ, written $\gamma[A]$, is the set of all points of X' of the form $\gamma(p)$, with p in A.

Definition. Let X and X' be sets, and $X \xrightarrow{\gamma} X'$. Let $A' \subset X'$. Then the *inverse image* of A' under γ, written $\gamma^{-1}[A']$, is the set of all points p of X with $\gamma(p)$ in the set A'.

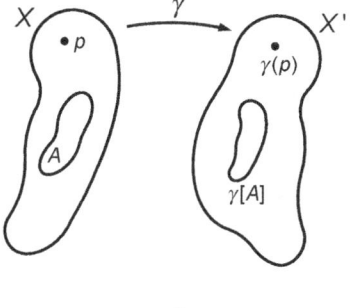

The first definition is illustrated in the figure on the right. We have set X and X', and rule γ which assigns, to each point of X, a point of X'. further we are given some subset A *of X*. Now γ assigns, to *each* point of X a point of X'. In particular, we may consider the points of A (which of course are also in X, since $A \subset X$).

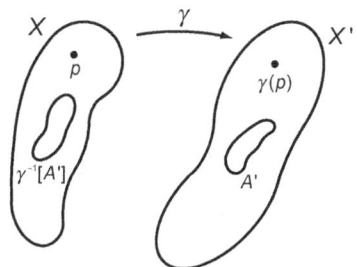

Given any such point, p, in A, we can apply our rule, and obtain a point, $\gamma(p)$, of X'. Let us now do this – apply γ – for *every* point of the set A. The result will be some points of X'. The set of all points of X' so obtained is the subset $\gamma[A]$ of X'. Thus, *one only takes the image of a subset of X, and the result, $\gamma[A]$, is always a subset of X'.* [Essentially, whereas the mapping starts out just sending each *point* of X to a *point* of X', consideration of the notion of "image" forces the mapping also to send each *subset* of X to a *subset* of X'.]

The second definition is illustrated in the second figure on the right above. Again, we have sets X and X', and mapping γ. Now, however, we consider subset A' *of* X'. Now γ assigns, to each point of X, a point of X'. Some of the resulting "points of X'" may happen to wind up in the subset A', and some may not. We now consider all points *of* X which the rule γ *does* send into the subset A' of X'. This subset of X is the inverse image, $\gamma^{-1}[A']$. Thus, *one only takes the inverse image of a subset of X', and the result, $\gamma^{-1}[A']$, is always a subset of X*. It is just the reverse of the "image". [It is easy to remember all this from the terminology. The image sends subsets of X to subsets of X', just as the original mapping sends points of X to points of X'. The inverse image is the reverse: It sends subsets of X' to subsets of X. We give some examples.

Let X be the set of numbers, X' the plane, and $X \xrightarrow{\gamma} X'$ a curve. Then $\gamma[X]$, the image X under this mapping, is the set of all points of the plane of the form $\gamma(t)$ for some number t. This is just "the set of all points of the plane through which the curve passes", i.e., the "points marked when we draw the curve".

Let X and X' both be the set of real numbers, and $X \xrightarrow{\gamma} X'$ the function of one real variable given by $\gamma(x) = x^2$. [Note: indeed a mapping.] Let A be the subset of X consisting of all numbers between 2 and 5. Then $\gamma[A[$, the image of A, is the set of all numbers (points of X') of the form $\gamma(x)$ for x between 2 and 5, i.e., the set of all numbers obtained by squaring a number between 2 and 5, i.e., the set of all numbers between 4 and 25. Next, let A' be the subset of X' consisting of all numbers between 4 and 25. Then $\gamma^{-1}[A']$ is the set of all numbers (points of X) such that $\gamma(x)$ is between 4 and 25. Hence, $\gamma^{-1}[A']$ consists of all numbers between 2 and 5, *and also* all numbers between (-2) and (-5) (for these also have squares between 4 and 25.

Let X be any set, and $X \xrightarrow{\gamma} X$ the identity mapping, with $\gamma(p) = p$ for every p in X. Then, for A a subset of X, $\gamma[A] = A$ (for the set of all points of the form $\gamma(p)$, with p in A, is just A, since $\gamma(p) = p$), and $\gamma^{-1}[A] = A$ (for the set of all points p with $\gamma(p)$ in A is just A).

Let X be any set, X' the plane, and $X \xrightarrow{\gamma} X'$ be the mapping with $\gamma(p) =$ origin, for any point p of X. Let A be any subset of X not the empty set. Then $\gamma[A]$ (set of points of X', the plane, one gets by applying γ to points of A) is the subset of X' consisting of just the origin. [Of course, γ [empty set = empty set.] Let A' be a subset of X' not including the origin. Then $\gamma^{-1}[A']$ (all points of X with $\gamma(p)$ in A' is the empty set (for $\gamma(p)$ is always the origin, while the origin is not in A'. So, $\gamma(p)$ is *never* in A', for any p.) For A' a subset of the plane X' including the origin, then $\gamma^{-1}[A']$ is all of X (for every point p of X has $\gamma(p)$ in A', since $\gamma(p)$ is the origin, and that is in A').

For the first example of two pages ago, γ^{-1}[yes] is just $\mathcal{J}$, the subset of X consisting of all open subsets of Y.

We can of course consider in particular mappings of topological spaces.

That is: Let $X, \mathcal{T}$ and $X', \mathcal{T}$ and $X', \mathcal{T}'$ be topological spaces, and $X \xrightarrow{\gamma} X'$. Our next task is to define a continuous mapping of topological spaces. The plan, of course, is to use as guides the definition of a continuous curve, the observation that what was really involved in that definition was a *metric space* of numbers and really a *metric space* of the plane, and the fact that every metric space gives rise to its underlying topological space. What we want to do, in short, is define a continuous mapping of topological spaces in a way so that the result is reminiscent of the definition of a continuous curve.

So, let X, d and X', d' be metric spaces, and $X \xrightarrow{\gamma} X'$. [To be more concrete, one could let X, d be the metric space of numbers, with $d(t, t_0) = |t - t_0|$, and X', d' the plane with geometrical distance.] Fix a point p of X, and suppose that γ is continuous at p. That is, suppose that, for any positive number ϵ, there exists a positive number δ such that, whenever $d(p, q) \leq \delta$ (where q, of course, is in X), $d'(\gamma(p), \gamma(q)) \leq \epsilon$. [Note that this last makes sense: $\gamma(p)$ and $\gamma(q)$ are points of X', and d' gives distances between points of X'. But "$d'(p, q)$", for example, makes no sense, for d' does not give distances between points of X.] The plan is to eliminate all mention of "distance", in favor of "self-interior sets" in our metric spaces. The resulting statement, using just "self-interior sets", will then suggest a definition in topological spaces, replacing "self-interior" by "open".

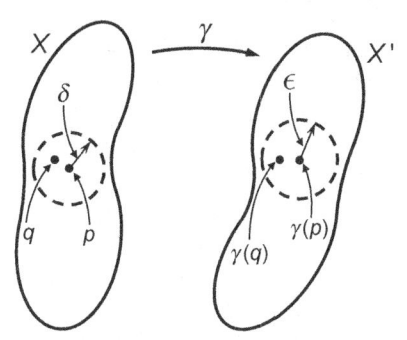

So, we had better start by considering a self-interior set somewhere, i.e., in either X, d or in X', d'. Choosing one in X, d as it turns out, does not get one very far. [Of course, in practice, if one were trying to invent "continuous mapping of topological spaces" for the first time, one would have to try everything in sight. It might take one a full week to come to the realization that starting with a self-interior set in X, d is not very promising!] In any case, we decide to consider self-interior sets in X'. Since we are dealing with continuity at the point p of X, our self-interior set in X', d' had better have something to do with p. [We cannot, of course, just demand that p be in this self-interior set, for p is in X, while our self-interior set is in X'.] The next best thing would be to choose a self-interior set C' in X' such that (p) (a point *of X'*) is in C'.

So, we so far have metric spaces X, d and X', d', point p of X, mapping $X \xrightarrow{\gamma} X'$ continuous at p (as above), and self-interior $C' \subset X'$ with $\gamma(p)$ in C'. Now, since C' is self-interior, we have $C' = \text{int}(C')$. So, since $\gamma(p)$ is in C', it is also in int (C'). That is to say, there exists a positive number ϵ such that every point q' of X' with $d'(\gamma(p), q') \leq \epsilon$ is in C'.

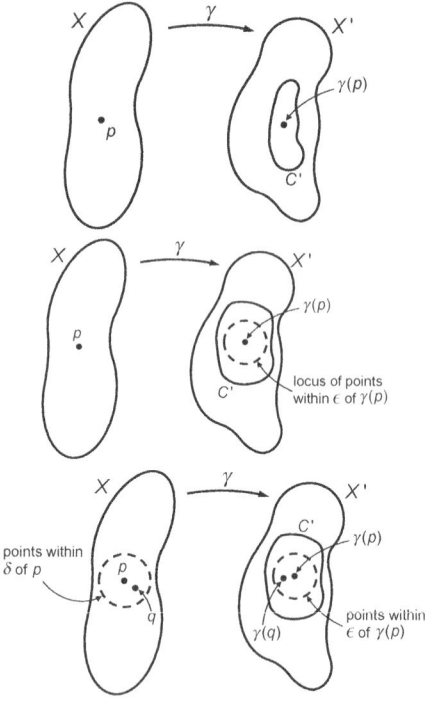

Now things are beginning to look promising. Recall the definition of continuous (on the left of the figure on the previous page): For any positive ϵ, there exists a positive δ such that, whenever $d(p,q) \leq delta$, $d'(\gamma(p), \gamma(q)) \leq \epsilon$. This must be true for *every* positive ϵ – and in particular for the ϵ we just obtained above (the one so that every point of X' within ϵ of $\gamma(p)$ is in C'). Thus, by continuity: There exists a positive number δ such that, for every point q of X with $d(p,q) \leq \delta$, $\gamma(q)$ is in C'.

Great! So far, by introducing a self-interior set in X', we have managed to eliminate all mention of the number ϵ. Next, we must eliminate mention of the number δ – and of course it must have something to do with self-interior sets in X'. How shall we do this? The key is to use the idea of inverse-image.

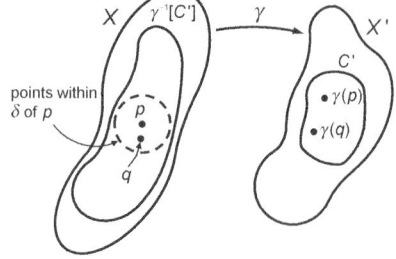

Just for fun, let us consider the inverse image of C', $\gamma^{-1}[C']$, i.e., the set of all points q of X with $\gamma(q)$ in C'. What can we say about this $\gamma^{-1}[C']$? Well, first of all, p is certainly in $\gamma^{-1}[C']$, for $\gamma(p)$ is in C' (for that is how we choose C' in the first place). What else must be in $\gamma^{-1}[C']$? The last sentence of the previous paragraph includes: "There exists a positive number δ such that, for every point q of X with $d(p,q) \leq \delta$, $\gamma(q)$ is in C'." In other words, using the definition of $\gamma^{-1}[C']$, "There exists a positive number δ such that every point q with $d(p,q) \leq delta$ is in $\gamma^{-1}[C']$." But look at this last statement: It just says that p is in int $(\gamma^{-1}[C'])$. We have now eliminated mention of the number δ!

So far, then, we have the following. We have metric spaces X, d and X', d', point p of X, mapping $X \xrightarrow{\gamma} X'$ continuous at p, and self-interior subset C' of X' with $\gamma(p)$ a point of C'. We conclude under these circumstances that p must be in the interior of $\gamma^{-1}[C']$. This conclusion, of course, just expresses the idea that our mapping γ (of metric spaces) is continuous at p. But it has been cleverly arranged so that it uses words such as "self-interior", "inverse image", and "interior", and not the word "distance".

For the final step, we have to work on the point p. Let us now suppose that $X \xrightarrow{\gamma} X'$ is continuous, not only at the single point p, but even at *every* point of X. [Just as a continuous curve requires continuity at every point.] Let us keep fixed the self-interior set C' and consider changing our choice of the point p of X. We wish to invoke our conclusion above, "p must be in the interior of $\gamma^{-1}[C']$". For which p's will this conclusion follow? Well, all the conditions were listed above: "... metric spaces X, d and X', d', point p of X, mapping $X \xrightarrow{\gamma} X'$ continuous at p, and self-interior subset C' of X' with $\gamma(p)$ a point of C'." What of all this involves point p? First, the mapping γ must be continuous at p: But this is ok, for we have just assumed the mapping continuous at *every* point of X. Second, the point p must be such that $\gamma(p)$ is in C'. So, this is the only condition on p. That is to say, we have that p must be in the interior of $\gamma^{-1}[C']$ *whenever* $\gamma(p)$ is in C'. But "$\gamma(p)$ is in C'" is the same as "p is in $\gamma^{-1}[C']$". So, we have that "every point p of $\gamma^{-1}[C']$ must be in the interior of $\gamma^{-1}[C']$". But look again at this last quote. [Of course, every point of the interior of $\gamma^{-1}[C']$ is in $\gamma^{-1}[C']$, by an earlier theorem.] It just says that $\gamma^{-1}[C']$ is self-interior! Thus, not only do we eliminate the number δ, but everything reduces finally to talking about self-interior sets.

What we have shown, then, is the following. Let X, d and X', d' be metric spaces, and $X \xrightarrow{\gamma} X'$. Then, if γ is continuous at every point of X [that is, if for every point p of X and every positive number ϵ there exists a positive number δ such that, whenever $d(p, q) \leq \delta$, $d(\gamma(p), \gamma(q)) \leq \epsilon$ – what a mess!], it follows that the inverse image under γ of every self-interior subset of X' is a self-interior subset of X. By just going trough the discussion above, one convinces oneself that the converse is also true, i.e., that if inverse images of self-interior sets are self-interior, then γ is continuous (i.e., all that $\epsilon - -\delta$ business holds). In short, we have succeeded in expressing the very messy idea of continuity for mappings of metric spaces just in terms of self-interior sets. The resulting expression is extremely simple: no "you choose this and I choose that", no complicated business about getting the order of choices right. All one has to do is take inverse images of self-interior sets in X', and check to see if the resulting subsets of X are also self-interior.

The last step is very easy. We just generalize from metric spaces to topological spaces by replacing "self-interior" by "open". All this suggests, then, the following.

<u>Definition</u>. Let $X, \mathcal{J}$ and $X', \mathcal{J}'$ be topological spaces, and $X \xrightarrow{\gamma} X'$. Then γ is said to be *continuous* if the inverse images under γ, of all open subsets of X' are open subsets of X.

The hope, then, is that this definition captures the idea that the mapping sends, "nearby" points of X, to "nearby" points of X' i.e., that the point $\gamma(p)$ of X' does not "skip" or "jump around" in X' when p moves just a little bit in X. As we all shall see, it does capture this idea.

We give some examples.

Let X, $\mathcal{J}$ be any topological space, and $X \xrightarrow{\gamma} X$ the identity mapping. Then, for any subset A of X (the one of the right), $\gamma^{-1}[A] = A$ (a subset of the X on the left). Obviously, inverse images of open sets are open. This mapping of topological spaces is continuous. [Intuitively, $\gamma(p)$ move in the X on the right just as p moves in the X on the left. There is no "jumping around".]

Let each of X, $\mathcal{J}$ and X', $\mathcal{J}'$ be the topological space of real numbers (i.e., the underlying topological space of the metric space of numbers, with distance $|t - t_0|$.) Let $X \xrightarrow{\gamma} X'$ be the following: Given a point t of X (i.e., a real number), let $\gamma(t)$ be 7 (a point of X') if $t < 0$, and 5 if $t \geq 0$. This is a mapping. [Intuitively, $\gamma(t)$ "jumps" from the point 7 of X' to the point 5 as t increases through 0. One would not expect it to be continuous.] This mapping is not continuous.

Let A' be the subset of X' consisting of all numbers x with $41/2 < x, 51/2$. This set is self-interior in the metric space, and so is open in the underlying topological space. Now $\gamma^{-1}[A']$ is the set of all numbers t (in X) with $41/2 < \gamma(t) < 51/2$. But, from γ is, this is just the set of all numbers t in (X) with $t \geq 0$. But this $\gamma^{-1}[A']$ is not a self-interior subset of X (for the number 0 is in this set, but not in its interior). That is, $\gamma^{-1}[A']$ is not open in the topological space X, $\mathcal{J}$. So, we have found an open set in X', $\mathcal{J}'$ whose inverse image is not open in X, $\mathcal{J}$. So, this mapping is not continuous.

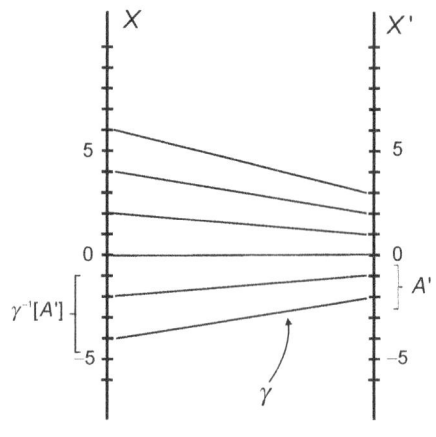

Let X, $\mathcal{J}$ and X', $\mathcal{J}'$ be as above, but let $X \xrightarrow{\gamma} X'$ be the mapping with $\gamma(t) = 1/2t$. [That is, this is the mapping "divide by two". Intuitively, division by two should take nearby numbers to nearby numbers. So, we expect this mapping to be continuous.] This mapping is continuous. Let A' be any open subset in X', $\mathcal{T}'$. Then $\gamma^{-1}[A']$ is just A' "blown up by a factor of two", i.e., all numbers of the form $2x$, for x in A'. This $\gamma^{-1}[A']$ must be a self-interior subset of the metric space. [Indeed, given a point, $2x$, of $\gamma^{-1}[A']$, find the ϵ which shows that x is in int (A'). Use 2ϵ to show that $2x$ is in int $(\gamma^{-1}[A'])$.] Thus, the inverse image of any open set in X', $\mathcal{T}'$ is open in X, $\mathcal{T}$.

Let X, $\mathcal{T}$ and X', $\mathcal{T}'$ be topological spaces. Fix once and for all a point p' of X'. Let $X \xrightarrow{\gamma} X'$ be the mapping with $\gamma(q) = p'$, for any q in X. [Intuitively, "all of X is sent to the single point p' of X'". There could hardly be any jumping around of $\gamma(q)$, since it is always the point p'. So, one expects continuous.] This mapping is continuous. Let A' be an open subset of X', with p' not in A'. Then $\gamma^{-1}[A']$ is the empty subset of X – an open set, by the first condition for a topological space. Let B' be an open subset of X', with p' in B'. Then $\gamma^{-1}[B']$ is all of X – an open set, by the first condition for a topological space. Thus, inverse images of open sets are open.

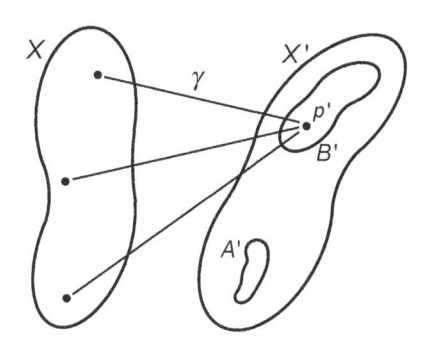

Let X, $\mathcal{T}$ be a discrete topological space, X', $\mathcal{T}'$ a topological space, and $X \xrightarrow{\gamma} X'$. Then this mapping is always continuous. Let A' be an open set in X', $\mathcal{T}'$. Then $\gamma^{-1}[A']$ is certainly *some* subset of X. But X, $\mathcal{T}$ is discrete, and so this is an open subset of X (for every subset of X is open in this topology). So, inverse images of open sets are open. [Intuitively, think of X, $\mathcal{T}$ as coming from the metric space, with $d(p,q) = 0$ if $p = q$, and 1 otherwise. Continuity is to ask whether "small changes" in the point p result in "large changes in $\gamma(p)$. But the points "p" are to be in this metric space. The only way one can get a "small change" from p (at least, if this "small change" is motion by a distance less than 1) is just to remain at p. Any other "change" of the point p moves one a whole distance of 1. So, of one only moves a "little bit" from p, i.e., if one stays at p, then $\gamma(p)$ will not change at all (because p could not change). So, indeed, "small changes" in the point p *do* result in small changes in $\gamma(p)$. So, one would expect continuous.]

Let X, $\mathcal{T}$ be a topological space, X', $\mathcal{T}'$ an indiscrete topological space, and $X \xrightarrow{\gamma} X'$. Then this mapping is always continuous. Let A' be an open subset of X'. Then, since X', $\mathcal{T}'$ is indiscrete, A' must be either the empty subset of X', or X' itself. For A' empty, $\gamma^{-1}[A']$ is the empty subset of X; for $A' = X'$, $\gamma^{-1}[A']$ is all of X. But, by the first condition for a topological space, both of these are open subsets of X. [Intuitively, any change of point in an indiscrete topological space is "small": The open sets are so insensitive to the points that can hardly tell one point from another. Since the open sets can not really tell the points apart, they regard any change of point as "small". So, when p chances a little bit, $\gamma(p)$ always changes a little bit, for the latter is a "change" in X', $\mathcal{T}'$, where everything is a "little bit". The indiscrete topological spaces are like "metric spaces" in which all distances are zero (although, of course, these are not metric spaces). If all distances are zero, every change of point is always by a "small distance".]

Every continuous curve defines a continuous mapping on the underlying topological spaces.

Let X, d be any metric space, and fix a point p of X. Let X' be the set of real numbers, and $X \xrightarrow{\gamma} X'$ the mapping with, for any point q of X, $\gamma(q) = d(p, q)$, a point of X' (i.e., a number). [Intuitively, the mapping is "evaluate the distance from p'". One would expect that "nearby points in X will have nearly equal distances from p". That is, one would expect the mapping to be continuous.] This mapping is continuous. We sketch the argument. Let A' be self-interior in the metric space of real numbers;: we must show that $\gamma^{-1}[A']$ is self-interior in X. Let q be any point of $\gamma^{-1}[A']$, i.e., let $\gamma(q) = a$ (a real number, namely $d(p, q)$) be in A'. We must show that q is in the interior of $\gamma^{-1}[A']$. That is, we must find a positive number δ such that every point of X within δ of q is in $\gamma^{-1}[A']$. That is, we must find positive δ such that every point r within δ of q has $\gamma(r)$ in A'. Since A' is self-interior (in X'), there is a positive number ϵ such that every number within ϵ of $\gamma(q) = a$ is in A'. Set $\delta = \epsilon$. Let r be any point within $\delta = \epsilon$ of q. Then by the triangle inequality, $d(p, q) + d(q, r) \geq d(p, r)$, or $a + \epsilon \geq d(p, r)$; and $d(p, r) + d(r, q) \geq d(p, q)$, or $d(p, r) + \epsilon \geq a$, or $d(p, r) \geq a - \epsilon$. But these two together say that $d(p, r) \leq a + \epsilon$, and $d(p, r) \geq a - \epsilon$. That is, they say that $d(p, r)$ is within ϵ of a. That is, they say that $\gamma(r)$ is within ϵ of a. But every number within ϵ of a is in A' (for that is how we choose ϵ). So, $\gamma(r)$ must be in A'. Thus, we have found our positive number δ such that every point r within δ of q has $\gamma(r)$ in A' (namely, $\delta = \epsilon$). So, inverse images of self-interior sets are self-interior. So, in the underlying topological spaces, the mapping is continuous.

I hope these examples at least suggest that the definition of "continuous" is a reasonable candidate for saying what we want to say. One strong piece of evidence that "topological space" is the right thing to be considering is that "continuous mapping" becomes so simple an idea in its framework. As the last example illustrates, a very simple definition can become quite complicated when it is actually applied.

14. Applications of Continuous Mappings

Many of the more interesting results in topology involve the interaction between continuous mappings on the one hand, and properties of various subsets of topological spaces on the other. In this section, we give some examples.

Let X, $\mathcal{T}$ and X', $\mathcal{T}'$ be topological spaces, and $X \xrightarrow{\gamma} X'$ a continuous mapping. Intuitively, γ "sends nearby points of X to nearby points of X'", i.e., it "sends X to X' without tearing". Now let A be a connected subset of X. Intuitively, A "consists of just one piece in X". In terms of these intuitive characterizations, one might expect that "since A is just one piece, and since no tearing is allowed under the mapping, the corresponding image set in X' must also consist of just one piece" – for, after all, this would seem to describe what we mean by "no tearing". These remarks suggest a result in topology.

Theorem. Let X, $\mathcal{T}$ and X', $\mathcal{T}'$ be topological spaces, $X \xrightarrow{\gamma} X'$ a continuous mapping, and A a connected subset of X. Then $\gamma[A]$ is a connected subset of X'.

Proof: Suppose, for contradiction, that $\gamma[A]$ were disconnected, so there is a subset B', of X' such that some point, p', of $\gamma[A]$ is in B', some point q', of $\gamma[A]$ is in B', some point, q' of $\gamma[A]$ is not in B', and no point of $\gamma[A]$ is in bnd B'). Set $B = \gamma^{-1}[\text{int}(B')]$, a subset of X.

We shall show that i) some point of A is in B, ii) some point of A is not in B, and iii) no point of A is in bnd (B). i) Since p' is in $\gamma[A]$, there is a point p of A with $p' = \gamma(p)$. Since p' is in B' and not bnd (B'), p' is in int (B'). So, since $\gamma(p) = p'$, p is in $\gamma^{-1}[\text{int}(B')] = B$. So, p is in A and B. ii) Since q' is in $\gamma[A]$, there is a point q of A with $q' = \gamma(q)$. But $\gamma(q)$ is not in B', and so not in int (B'), and therefore q cannot be in $\gamma^{-1}[\text{int}(B')] = B$. So, q is in A and not in B. iii) Suppose, for contradiction, that there were a point r in A and bnd (B), Then $\gamma(r)$ is in $\gamma[A]$, and so, since no point of $\gamma[A]$ can be in bnd (B'), $\gamma(r)$ must be in either int (B') or int (B'^C). Were $\gamma(r)$ in int (B'), then r would be in $\gamma^{-1}[\text{int}(B')] = B$. But int (B') is open in X', $\mathcal{T}'$, and so, by continuity of γ, B is open in X, $\mathcal{J}$, and so $B = \text{int}(B)$. So, since r is in B, r is in int (B) – and hence r cannot be in bnd (B). Were $\gamma(r)$ in int (B'^C), r would be in $C = \gamma^{-1}[\text{int}(B'^C)]$. But int (B'^C) is open in X', $\mathcal{J}'$, and so, by continuity of γ, C is open in X, $\mathcal{J}$. So, C is an open neighborhood of r. But r is in bnd (B), and so, since C is an open neighborhood of r, there must be a point u in both C and B. Since u is in C, $\gamma(u)$ must be in int (B'^C), while, since u is in B, $\gamma(u)$ must be in int (B'). But no point of X' can be in both int (B'^C) and int (B'). We conclude, then, that the existence of a point r in both A and bnd (B) leads to a contradiction. Hence, there is no point of A in bnd (B).

We supposed that $\gamma[A]$ were disconnected, and found a set B such that some point of A is in B, some point of A is not in B, and no point of A is in bnd (B), That is, we found A disconnected, But this contradicts the hypothesis.

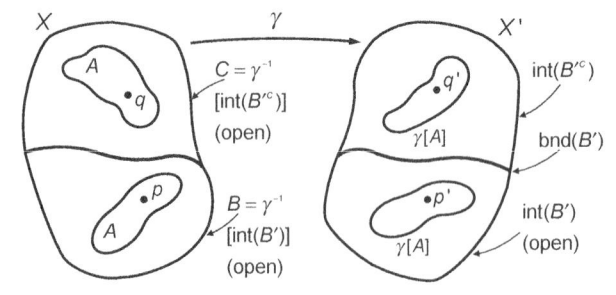

This is of course quite a complicated proof. We have connected $A \subset X$, and we suppose that $\gamma[A]$ is disconnected, in order to obtain a contradiction. We consider the set B' which "disconnects" $\gamma[A]$. The idea is to use this B' to show that A is disconnected, i.e., to arrive at a contradiction. This is done by taking a lot of inverse images. The "two "pieces" into which $\gamma[A]$ is separated are represented by int (B), and int (B'^C). [All that remains is bnd (B') – but no points of $\gamma[A]$ are in there.] These are open sets in X', and so, by continuity, their inverse images, called B and C above, are open sets in X. These are the two open sets which will "separate A into two pieces". That B actually has the three properties necessary to "disconnect" A (some point of A in B, some point of A not in B, and no point of A in bnd (B)) is shown by using for each, the corresponding property for B' "disconnecting" $\gamma[A]$. We essentially "take the inverse image of each property". The first two are easy, while the third, iii) above, is a bit tricky – using, for example, the definition of boundary.

The thrust of the theorem is that, again, the actual definitions of "continuous" and "connected" seem to reflect our intuition.

We give two examples of more "concrete" applications of this theorem.

Consider ordinary real-valued function of one real variable. Such a function can be represented by its graph (of y as a function of x, say). Continuity for such a function means, roughly speaking, that the graph is "one continuous line, with no jumps". It seems "obvious" that, if a *continuous* function somewhere becomes positive (i.e., $f(a)$ is positive for some number a) and somewhere becomes negative (i.e., $f(b)$ is negative for some number b), then that function somewhere is zero (i.e. $f(c) = 0$ for some number c). This is true.

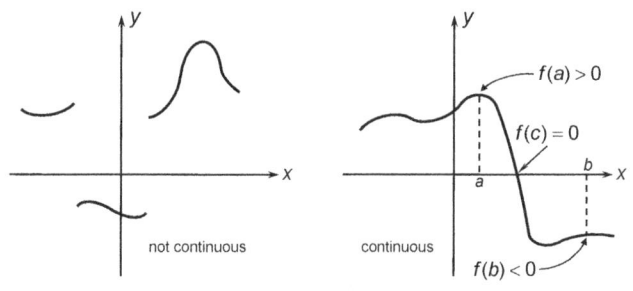

Theorem. Let each of X, $\mathcal{J}$ and X', $\mathcal{J}'$ be the usual topological space of real numbers, and $X \xrightarrow{f} X'$ a continuous mapping. Let, for some a in X, $f(a) > 0$, and, for some b in X, $f(b) < 0$. Then, for some c in X, $f(c) = 0$.

Proof: For contradiction, suppose not. Let $A = X$, a connected subset of X. Then, by the previous theorem, $f[A]$ is a connected subset of X'. Let $B' \subset X'$ be the set of all positive numbers. Then some point, $f(a)$, of $f[A]$ is in B', and some point, $f(b)$, of $f[A]$ is not B'. But bnd (B') consists of just the number 0, while we have supposed that for no c does $f(c) = 0$. So, no point of $f[A]$ is in bnd (B'). We conclude, thus, that $f[A]$ is disconnected, a contradiction.

That is, we consider the connected subset X of X. Its image must be connected. But this image contains a positive number, $f(a)$, and a negative number, $f(b)$ – so how could this image be connected unless it also contains 0? [The B' makes this explicit.] But $f[A]$ containing zero means exactly that $f(c) = 0$ for some c in X.

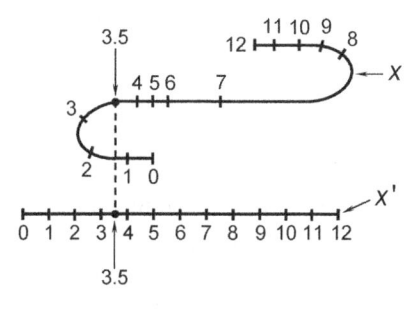

The second application of our theorem is one of a family of results called fixed-point theorems in topology. Imagine two rubber rules (whose markings run, say, say, from zero to twelve inches). Leave one ruler alone on the table, and take the other ruler, stretch or bend it in any way one wants (but no tearing!), and lay it next to the first ruler. This is to be done such that no part of the second ruler extends beyond the first ruler. A typical example of the result of this operation is shown in the figure. It turns out that, no matter how one stretches, bends, or folds (provided no spindling or mutilating) the second ruler, there will always exist some mark on the second ruler which winds up directly in the line with the corresponding mark on the first ruler! In the example above, for instance, the mark "3.5 inches" on the second ruler is in line with the mark "3.5 inches" on the first ruler. A few trials will convince one that this always works out this way (but it is not entirely obviously).

There is a theorem is topology which reflects this experiment. We replace the ruler left on the topological space. Let X' be the set of numbers x with $0 \leq x \leq 12$, and let the topology be the usual one. [That is, first make X' a metric space, with distance $d(x, x') = |x - x'|$, the usual distance between numbers. Then take the underlying topological space.] The ruler which is stretched and bent is represented by the same topological space, which we write $X, \mathcal{T}$. So, we so far have our rulers represented by topological spaces. The operation of "bending and stretching and laying next to" is just a thinly disguised mapping, $X \xrightarrow{\gamma} X'$, for this operation indeed assigns, to each point of the "bent and stretched" ruler, a point of the "left alone" ruler. That no tearing is allowed is expressed by demanding that this mapping be continuous. Finally, the conclusion we wish to draw is that some mark on the stretched ruler ends up directly in line with the corresponding mark on the stationary ruler. This is expressed by asking that $\gamma(a) = a$ for some point a of X (the stretched ruler). Thus, we have translated our "experiment" into mathematics. We have:

Theorem. Let $X, \mathcal{T}$ and X' and $\mathcal{T}'$ be the topological spaces above, and $X \xrightarrow{\gamma} X'$ continuous. Then, for some point a of X, $\gamma(a) = a$.

Proof: Suppose, for contradiction, that for no point a of X does $\gamma(a) = a$. Let X'' be the set consisting of just two points, p and q with the discrete topology. Let $X \xrightarrow{\mathcal{J}} X''$ be the following mapping: If $\gamma(a) > a$, set $\mathcal{J}(a) = p$; if $\gamma(a) < a$, set $\mathcal{J}(a) = q$. This is indeed a mapping, since, by supposition, $\gamma(a) = a$ for no a.

We show that $\mathcal{J}$ is continuous. It suffices to show that $\mathcal{J}^{-1}[p]$ and $\mathcal{J}^{-1}[q]$, subsets of X, are open. To this end, let a be a point of $\mathcal{J}^{-1}[p]$, so $\gamma(a) = a' > a$. Set $\epsilon = 1/2(a' - a)$, a positive number. Denote by B' the set of points x' of X' with $|a' - x'| < \epsilon$, so B' is open in $X', \mathcal{T}'$. By continuity of γ, $\gamma^{-1}[B'] = B$, the set of points x of X with $|\gamma(a) - \gamma(x)| < \epsilon$, is open in $X, \mathcal{T}$. Let C be the set of points x of X with $|a = x| < \epsilon$. Then C is also open in X. By the second condition for a topological space, $B \cap C$ is also open in X – and clearly a is a point of $B \cap C$. So, $B \cap C$ is an open neighborhood of a in X. Let b be any point of $B \cap C$. Then b is in B, and so $\gamma(b)$ is within ϵ of $\gamma(a)$; and b is in C, and so b is within ϵ of a. But $\gamma(a) = a + 2$, by our choice of ϵ. So, we have $\gamma(b) > b$. That is, b is in $\mathcal{J}^{-1}[p]$. Thus, for any point, a, of $\mathcal{J}^{-1}[p]$, we have found an open neighborhood, $B \cap C$, of a, such that any point of $B \cap C$ is in $\mathcal{J}^{-1}[p]$. That is, we have shown that $\mathcal{J}^{-1}[p]$ is open in $X, \mathcal{T}$. Similarly, $\mathcal{J}^{-1}[q]$ is open. So, $\mathcal{J}$ is continuous.

Since X is open connected and $\mathcal{J}$ is continuous, $\mathcal{J}[X]$, by the theorem on page 115, is connected in X''. But X'', $\mathcal{J}''$ is discrete, and so, in order that it be connected, $\mathcal{J}[X]$ must consist of just the point p or just the point q. Now certainly $\gamma(0) > 0$, so $\mathcal{J}(0) = p$, so p is in $\mathcal{J}[X]$; and $\gamma(12) < 12$, so $\mathcal{J}(12) = q$, so q is in $\mathcal{J}[X]$. So, both p and q are in $\mathcal{J}[X[$: a contradiction.

The crucial part of the proof is to show that, if $\gamma(a) > a$, then "sufficiently nearby" a''s have also $\gamma(a') > a'$. This, of course, is what one would expect, from continuity of the mapping γ. The business about ϵ is just to pin down "how sufficiently nearby" a' must be. Once we have this, we have that $\mathcal{J}^{-1}[p]$, and similarly $\mathcal{J}^{-1}[q]$, are open – so that $\mathcal{J}$ is a continuous mapping. We just "invented" X'' and $\mathcal{J}$ in order to be able to use the theorem on page 115. We now do so, to conclude that *both* p and q cannot be in $\mathcal{J}[X]$ – i.e,. that either $\gamma(a) > a$ for all a, or $\gamma(a) < a$ for all a. Finally, we obtain a contradiction, since $\gamma(0) > 0$ and $\gamma(12) < 12$. Note that, if we "leave the end points off the rulers", i.e., consider only x with $0 < x < 12$, then the proof above fails, in the last step. In fact, it is not difficult to find, for for "rulers without endpoints" a way to match up the two rulers so that no mark of X is aligned with the corresponding mark of X'.

We give a second example of an application of continuous mappings. Let X, $\mathcal{J}$ and X', $\mathcal{J}'$ be topological spaces, and $\xrightarrow{\gamma} X'$ continuous. Intuitively, γ "bends and stretches, but does not tear". Let A be a compact subset of X, so, intuitively, A is "well bounded in" in X. One might expect, then, that the image of A, $\gamma[A]$, must also be "well bounded in", for this could fail only if the action of γ is to "stretch A by an infinite amount, to make its image become unbounded", while a "stretching by an infinite amount" is practically a "tear". [This "argument", perhaps, is not a very convincing one. The problem seems to be that there is not a terribly good description, in non-technical words, of "compact".] In any case, such a result is true.

Theorem. Let X, $\mathcal{J}$ and X', $\mathcal{J}'$ be topological spaces, $X \xrightarrow{\gamma} X'$ continuous, and A a compact subset of X. Then $\gamma[A]$ is compact.

Proof: Suppose, for contradiction, that $\gamma[A]$ were not compact. Let the B''s be a collection of open sets which covers $\gamma[A]$, such that no finite number covers $\gamma[A]$. For each B' in this collection, set $B = \gamma^{-1}[B']$, so, by continuity of γ, each subset B of X is open. Foe any point p of A, $\gamma(p)$ is of course in $\gamma[A]$, and so $\gamma(p)$ is in one of the sets B' (since they cover $\gamma[A]$), and so p is in $B = \gamma^{-1}[B']$. Thus, every point of A is in one of the $\mathcal{J}B$'s i.e., the B's cover A. But A is compact, and so a finite number of the B', say $B_1, B_2, \ldots, B_n$ cover A. We claim that, as a consequence, the corresponding sets in X', $B'_1, \ldots, B'_n$, cover $\gamma[A]$. Indeed, let p' be any point of $\gamma[A]$, so $p' = \gamma(p)$ for some point p of A. Since $B_1, \ldots, B_n$ cover A, p must be in one of these, say B_1. But p in $B_1 = \gamma^{-1}[B''_1]$ means that $p' = \gamma(p)$ is in B'_1. So, this p' is in one of $B'_1, \ldots, B'_n$, namely B_1.

We conclude, then, that a finite number of the B''s cover $\gamma[A]$ – which contradicts our choice of the B''s.

The argument is rather like that on page 115. We just use inverse images to "pull back" things from X' to X. Continuity, of course, plays a central role, for it ensures that we do not ruin open sets when we "pull them back to X".

We give one example of an application of this theorem. Consider again real-valued functions of one real variable. We say that such a function "assumes its maximum" if there is a number a such that $f(a)$ is "at least as large as f of anything ever gets to be". For example, the function illustrated in the first graph on the right assumes its maximum – for $f(a)$, as indicated, is at least as large as any $f(b)$. The second function illustrated does not assume its maximum: In fact, it does not have any maximum at all, for $f(x)$ "just keeps getting bigger as x gets bigger". The third function illustrated doe not "keep getting bigger and bigger" – but still it does not assume its maximum. The problem here is that, even though the function itself never exceeds the value one, the function is so arranged that $f(a)$ is actually -1: The function makes a "jump" at a, to avoid having $f(a) = 1$. [Of course, were $f(a)$ one, then the function would assume its maximum.] Under what conditions might one expect a function to assume its maximum? The "reason" why the function in the last example does not assume its maximum is that there is a "jump" just where it ought to be assuming its maximum. We can prevent this possibility by demanding that the function be continuous. For the other example, the problem is that the function "trails up to large values for $f(x)$ as x trails off to infinity". We can prevent this sort of thing by just considering the function for x's in some "finite region", say $0 \le x \le 1$. Then, of course, we shall only be able to show that the function "assumes its maximum compared with other values it assumes for x with $0 \le x \le 1$".

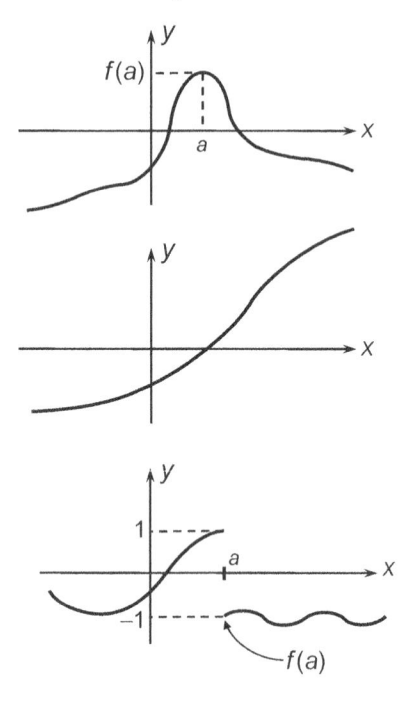

Are these safeguard sufficient to ensure that a function assume its maximum? The next theorem will show that they are. In the proof, we shall make use of two elementary facts about numbers: i) The set of all numbers x with $0 \le x \le 1$ is compact (virtually a consequence of the theorem on page 100), and ii) given any compact subset of the set of real numbers, there is an element of that subset which is larger than any other element of that subset.

Theorem. Let each of X, $\mathcal{J}$ and X', $\mathcal{J}'$ be the usual topological spaces of real numbers, and X', $\mathcal{J}'$ be the usual topological spaces of real numbers, and $X \xrightarrow{\gamma} X'$ a continuous mapping. Let $A \subset X$ be all points x of X with $0 \leq x \leq 1$. Then there exists a point a of A such that $f(b) \leq f(a)$ for any point b of A.

Proof: By the theorem above, since A is a compact subset of X, $f[A]$ is a compact subset of X'. So, there exists a point a' of $f[A]$ is such that every point b' of $f[A]$ satisfies $b' \leq a'$. Since a' is in $f[A]$, there is a point a of A with $f(a) = a'$. Then, for any point b of A, $f(b)$ is in $f[A]$, and so $f(b) \leq f(a)$.

These, of course, are just a few selected examples. They do not begin to do justice to the enormous range and complexity of result in this subject. I hope, however, that they do illustrate the facts that topology does make contact with more "concrete" things, and that the original program – to build this subject entirely on the definition of a topological space – is at least a promising one.

15. Conclusion

Topology is one of the branches of modern mathematics. It is of course difficult to divide any subject – and mathematics in particular – into neat, clearly defined, "branches". An attempt to do so in the case of mathematics might yield two others: algebra and analysis. Algebra is essentially the study of "operations", of the results of manipulating elements of sets according to certain, prescribed, rules. The objects with which one deals (analogous to topological spaces in topology) are such things as groups and vector spaces. If one had to say in a sentence the difference between algebra and topology, it would perhaps be that in algebra one is concerned primarily with individual elements of sets; in topology, with subsets. Analysis, on the other hand, involves functions (usually, of real or complex variables), derivatives, integrals, differential equations, and so on. It is, for example, the area in which calculus largely lies. What is usually called "algebra" in high school, however, is perhaps 30% algebra, and 70% analysis. Of the three branches, algebra and topology are perhaps are the most self-contained. Analysis often drawn heavily on the other two – and in particular on topology. There are, of course, many areas of mathematics which lie between these branches (for example, Lie groups, which are almost exactly halfway between algebra and topology, with a dash of analysis), and others (such as axiomatic set theory) which are separate from them all.

Mathematics finds application to the other sciences, and so topology, as a part of mathematics, itself finds application. While it would be impossible to describe in detail such applications here, a few examples – all to physics – may illustrate this idea.

Consider a box containing a gas. Denote by X the set of points of the box (so, for example, we might regard X as the set of points, (x, y, z) of Euclidean three-dimensional space, with $0 \leq x \leq 1$ $0 \leq y \leq 1$, and $0 \leq z \leq 1$). At each point of the box, one can determine the temperature, T, of the gas. Of course, different points could have different temperatures: The gas could be cold at the bottom of the box, and warm at the top. Thus, "measurement of the temperature" could be represented as a rule which assigns, to each point of X (point of the box), a number (the temperature at that point). We thus acquire a mapping from X to X', the set of real numbers. One knows in physics that heat flows from regions of high temperature to regions of low temperature. Furthermore, the faster the temperature changes (say, per inch of motion in the box), the faster the heat flows from one region to the other. A consequence of this is that one cannot have the temperature "change suddenly" between two nearby points of the box, for the flow of heat would immediately cool the hot region and warm the cold region. That is to say, the temperature, as a function of position in the box, cannot "jump". This would be reflected in the mathematics as follows. We would regard the set X as a topological space (the underlying one, from geometrical distance in Euclidean space), and the set X' as a topological space (the usual one). Then, we would demand that our mapping be continuous. Thus, a physical idea, about the way temperature behaves, would be represented as a mathematical one, about continuity of a mapping. Then, for example, the theorem on page 115 would say that, if two temperatures are achieved in the box, then every temperature in between is also achieved, somewhere in box. The theorem on page 119 would say that there is some "hottest point" in the box.

A mechanical system in physics is initially described by a set X which represents "the set of all possible configurations of the system", and is called the configuration space of the system. For example, consider the system of a ball free to roll on a table. A point of X would then be represented by giving the location of the ball on the table, and the origination of the ball in space. [In this example, the configuration space would be five-dimensional – two dimensions to tell where the ball is on the table, and three to tell the orientation of the ball.] One has a physical sense of what it means for two configurations to be "nearby". For example, two configurations, each having the ball at the same point of the table, but with one having the ball slightly rotated relative to the other, would be regarded as "nearby" – as would two configurations having the ball with the same orientation but at slightly different places on the table. This physical sense would be a reflected by regarding X as a topological space. Thus, as part of our physical ideas, the configuration space of a system is to be a topological space.

Now suppose that the table is tilted somewhat, so the ball, when released, will roll. Consider the experiment "begin with the ball at rest in some configuration, release the ball, and see what its configuration is five seconds later (after the ball has rolled a bit)". This is a rule which assigns, to each configuration (the initial one), some configuration (the final one). This would be represented, then, by a mapping from X to X. One aspect of the idea of dynamical stability in physics is that, if one changes very slightly the initial configuration of the ball, its ultimate configuration will not be changed very much, either. If this were not true, then the slightest little disturbance of the initial configuration would change dramatically what happens to the ball – something we just do not see happen in Nature. Thus, "dynamical stability" in physics would be reflected in our mathematical description by demanding that our mapping "with five seconds" be a continuous mapping of topological spaces.

But topology – as I hope has been made clear, – is not just a handmaiden of physics. It is an intellectual endeavor in its own right. It seeks to capture, within the framework which is mathematics, the notion of "closeness". The boundary of a set is the set of points which are "close to the set", and also "close to being out of that set"; a continuous mapping of topological spaces is one which preserves the "closeness" characterized by those spaces. This "captured closeness" – this notion of a topological space – has a sense of distance in it: Not the exact numerical values for distances, but rather enough of the shadow of distance to talk about what one wants to talk about.

Topology, at least for me, is a particularly pretty and aesthetic subject. Two aspects of the subject make it so. First, one puts so little into the definitions, yet so many tricky theorems come out. There just is not much to a topological space – just a set with some subsets satisfying a couple of simple properties. The various definitions – of connected, continuous mappings, and so on – are about as simple as definitions could be. But, on these few elements, one can build theorems which are very complicated indeed: for example, that the continuous image of a connected set is connected; that the attachment of the boundary to a connected set yields a connected set. These few theorems, of course, just scratch the surface. There are numerous theorems in this subject which are many times more complicated than these. [And, of course, we have presented only a small fraction of the definitions in topology.] The definitions build on each other, and intertwine, to produce a most appealing mosaic. The second remarkable thing about topology is that, not only does one have access to many complicated theorems, but furthermore these theorems on the whole even "make sense" intuitively. Surely there is no obvious reason why things work out this way. Look at the definition of a topological space. It certainly does not seem to scream out "Here is represented the essence of our ideas of 'closeness'." But that is exactly what is represented! How does it happen that such simple definitions, by means of complicated theorems which intertwine those definitions, turn out to mean something to our geometrical intuition? This is the sort of thing one usually wants in mathematics – and it is exactly what happens in topology.

But topology has also played a second role: It has served as an example for all of mathematics. Mathematics, as we have seen in this example, proceeds on two levels. On the one hand, there is the intuitive level, that in which one gets ideas, invents definitions which say what they are supposed to say, decides what statements are likely to be theorems and what are not, decides what line is likely to produce a proof of a theorem. The other level is that of precision. One proves theorems. Here on just uses the *statements* of the definition, the hypotheses of the theorems, and so on, not what those things are as ideas. One mechanically follows the rules, and produces proofs. Herein, at least for me, lies much of the charm of mathematics. It is almost as though there are two entirely different subjects at work here. From each, one only dimly sees the other. But, despite all this, they end up saying virtually the same thing. The second level – the precise one – is of course the more difficult for the beginner. One is used to making great leaps, of getting to the answer, to using every thought one can generate. In constructing proofs, however, one must hold oneself back. One must simply follow the rules (which one has made for oneself!) and go where one is led.

So, when one begins mathematics, there is a hard part (writing proper proofs), and an easy part (getting ideas, guessing what is true and what false). But, as one goes on, the roles reverse. One soon becomes accustomed to writing out correct proofs. it becomes mechanical, once one has seen the idea of the proof. The hard part – the stuff of much research in mathematics – becomes thinking of new definitions which "say" new things, finding new and unexpected relationships between old definitions, and even finding counterexamples to show that "widely expected" relationships simply do not hold.

The style of mathematics is not well-adapted to everything. It would be a bit silly, for example, to begin psychology with "Definition. A *person* is a ...", and later obtain "Theorem. There are exactly two sexes." For those things to which it *is* adapted, however, it delivers certain benefits. The definitions and theorems in mathematics "live forever". They are not to be changed by "subsequent evidence". [But, of course, there are styles in mathematics as in everything. Some definitions or theorems may later be regarded as uninteresting because of change of taste.] There is a kind of permanence in mathematics one does not see in other intellectual endeavors. Mathematics gives a new meaning to the word "true". But all this is at a price. Often, it is that one has to work very hard to show something which is "obvious". For example, "if one attaches the boundary to a connected set, the result is connected." seems clearly, under any reasonable interpretations of the words, "true". But we must, in mathematics, work comparatively hard for the proof. Sometimes all this work pays off. Something which seemed clearly to be true is just plain false. It usually means that we did not understand intuitively, in the fullest sense, the nature of the objects with which we were (or thought we were) dealing. More often, the work does not pay off so directly. It is just the price one pays for being able to say "true" as a mathematician.

Mathematics is not to everyone's taste, any more than any other intellectual endeavor. Why, if one does not care for it, should one know anything about it? I think that there may be something of value in mathematics for almost everyone. I do not mean in balancing the checkbook: That is hardly "mathematics". Mathematics forces one to a way of thinking very different from that in which one is accustomed. Perhaps a flexible mind is a viable end in itself. Stretching exercises, unfortunately, are not easy to come by. One might, for example, search for societies out of contact with Western civilization, join such a society for a few years, and come to understand how the minds of the members work. We are rapidly running out of these. Fortunately, mathematics, while perhaps not entirely "out of contact with western civilization", can serve. In short, I feel that mathematics can be useful because it is different in a very real sense – and different it certainly is.

Finally, I hope that I have managed to convey the fact that there is such a thing as creativity in mathematics – and such a thing as excitement.

15.

Appendix

Phy Sci 112 February 17, 1978

Is the Curve Continuous?

Rules: This is a game for two players, Yes and and No. The two players begin by agreeing on a curve γ (i.e., a rule which assign, to each number t, a point of the plane), and also on a number t_0. The game itself consists of three moves: 1. No announces a positive number ϵ. 2. Yes announces a positive number δ. 3. No announces a number t such that $|t_0 - t| \leq \delta$. At the completion of the three moves, the players together determine (using the rule γ) the points $\gamma(t_0)$ and $\gamma(t)$ of the plane, and then the number $d(\gamma(t_0), \gamma(t))$. If $d(\gamma(t_0), \gamma(t)) \leq \epsilon$, then Yes wins the game. Otherwise, No wins.

Strategy: No will win at the end if it is not true that $d(\gamma(t_0), \gamma(t)) \leq \epsilon$. No wants, with his first move, to make this as unlikely as possible. He does so by choosing ϵ to be a small positive number. Yes will win at the end if it is true that $d(\gamma(t_0), \gamma(t)) \leq \epsilon$. Yes, of course, can do nothing about ϵ; but he does control δ in his second move. The role of δ is to restrict the range of No's choices of t in the third move. Yes wants to restrict No as much as possible, and so he chooses δ to be a small positive number. No, for his third move, already knows ϵ and δ. So, in order to win, he must find t with $|t_0 - t| \leq \delta$, but not $d(\gamma(t_0), \gamma(t)) \leq \epsilon$.

General Remarks: This is of course a game of pure skill: There is no luck. Either player can certainly blow the game by using poor strategy. [For example, No could lose a winning game in the third move by selecting a poor t; Yes could choose δ too large in the second step, and thus give No too many possibilities for the t he is to choose in the third move.] Suppose, however, that both players are experienced, and use the best possible strategy. Then, clearly, the winner is already predetermined by the choice of the curve γ and the number t_0. Problem: State what must be true of γ and t_0 in order that Yes, using prefect strategy, can always win the game. Answer: Your statement will be exactly the definition of a curve continuous at t_0.

Supplementary Problems   Due: January 27, 1978

[Not to be turned in. Caution: These problems were "test rejects". In some cases, they were rejected because they were too ambiguous, or because they were too hard.]

1. Think of a set in the plane as "discrete" if every point of A is "separate from" the rest of A; no point of A is "intermingled with" the rest of A. Invent a definition of "discrete".

2. Give a formal proof that the boundary of a bounded set is bounded.

3. Is it true that, for any sets A and B in the plane, bnd $(A \cup B)$ is a subset of bnd $(A) \cup$ bnd (B)?

4. Let A, B, and C be connected sets in the plane, and let A and B have a point in common and B and C a point in common. Does it follow that the union of all three sets is connected?

5. Is it always true that int $(\text{bnd}(A)^C) = \text{bnd}(A)^C$?

6. Find sets A and B in the plane such that bnd $(A) = B$ and bnd $(B) = A$.

7. Is bnd (int (A)) $\subset$ bnd (A) for every set A in the plane?

8. One thinks of a set as disconnected if it "consists of two or more pieces". Try to find a definition of "one of those connected pieces of which the set consists". The definition would read: "For A a set in the plane, a subset of B of A is a *connected piece* of A if ..."

9. Let n be a positive integer, and let A be a set with exactly n points. For which values of n is A connected? For which values bounded? What is int (A)?

10. Let A be the set of all points (x, y) with $x^2 \neq y^2$. Find int (A) and bnd (A). Is A connected? bounded?

11. Find A such that neither A nor A^C is connected.

Phy Sc 112   **Test 1**   January 30, 1978

1. (12 points) Let A be the set in the plane consisting of all points (x, y) with $x^+y^2 < 1$ or $y = 0$, i.e., a "disk with a line running through it". Indicate in a clear diagram the sets bnd (A) and int (A). Give the formulae for these sets. Is A bounded? Is A bounded? Is A connected? [Please answer each of these last two in a sentence or two, with a reason.]

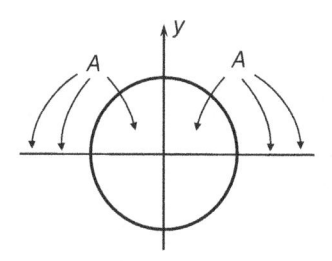

2. (20 points) Please answer in a sentence or two, with a reason.

i) Is it true that, for A any bounded set in the plane, int (A) is bounded?

ii) Is it true that, for any two sets in the plane having no point in common, their union is disconnected?

iii) Is it true that, for any set A in the plane with int (A) connected, A must be connected?

iv) Is it true that, for any set A in the plane, every point of A is in either int (A) or bnd (A)?

3. (10 points) <u>Theorem</u>. Let A be a set in the plane. Then no point of the plane is in both int (A) and bnd (A). [Give a formal proof.]

4. (10 points) <u>Definition</u>. A set A in the plane is said to be *closed* if, for every point p not in A, there exists a positive number ϵ such that no point q with $d(p, q) \leq \epsilon$ is in A. Is the usual disk (all (x, y) with $x^2 + y^2 < 1$) closed? Is the line (all (x, y) with $y = 1$ closed)? [Please answer in a sentence or two, with a reason.]

5. (8 points) Find a disconnected set A in the plane such that bnd (A) is the line by all (x, y) with $y = 1$.

Test 1 – Solutions

1. The sets bnd (A) and int (A) are as shown in the diagram. bnd (A) is the set of all points (x, y) with $x^2 + y^2 = 1$, or $y = 0$ and $x > 1$, or $y = 0$ and $x < -1$. int (A) is the set of all points (x, y) with $x^2 + y^2 < 1$. Set A is not bounded (and, indeed, the line itself is not even bounded): There is no point p of the plane and positive number c such that every point of A is within distance c of p. The set A is connected, because each of the line and disk by itself is connected, these sets have a point in common, and A is their union.

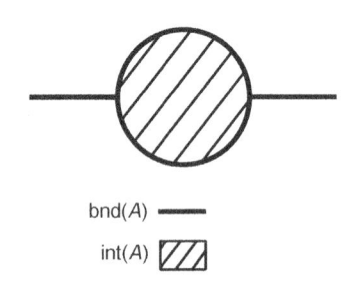

2. i) Yes. Int (A) is a subset of A, and every subset of bounded set is bounded. ii) No. Let A be be any set in the plane, and set $B = A^C$. Then A and B have no points in common, while their union, the entire plane, is connected. iii) No. Let, for example, A consists of the usual disk together with the point $(17, 17)$. Then A is certainly not connected, while int (A) is just the disk, which is connected. iv) Yes. Let p be a point of A. If for some positive ϵ every point within ϵ of p is in A, then p is in int (A). If this does not hold (i.e., if for every positive ϵ there is a point within ϵ of p and not in A), then, since of course there is for every positive ϵ a point, namely p, within ϵ of p and in A, p must be in bnd (A).

3. Proof: Let p be a point of int (A), so, for some positive number ϵ, every point within distance ϵ of p is in A. Then it cannot be the case that for every positive ϵ there is a point within ϵ of p and not in A. Hence, p cannot be in bnd (A).

4. The disk is not closed. The point $p = (1, 0)$ is not in A, and yet for every positive number ϵ there is a point within ϵ of p and in A. The line is closed. Given any point p not in A, let ϵ be one-half the perpendicular distance from p to the line, a positive number. Then it is indeed true that no point within distance ϵ of p is in A.

5. Let, for example, A be the complement of the line, i.e., the set of all points (x, y) with $y \neq 1$. Then A is disconnected, but its boundary is the line. [A second type of example is: Let A consist of all (x, y) with $y = 1$ and $x \neq 13$.]

Phy Sc 112   **Test 2**   February 20, 1978

1.(6 points) Give an example of a metric space in which every bounded set is connected.

2. (16 points) For each of the statements below, tell whether it is true or false, and explain in a few sentences:

   i) Let X, d be a metric space. Then int $(X) = X$.

   ii) Let X, d be a metric space, and $A \subset X$. Then, if A and A^C are both bounded, X is bounded.

   iii) Let X, d be a metric space, and $A \subset X$. Then if A and A^C are both connected, X is connected.

   iv. Let X, d be *any* metric space with X the plane. Let A be the usual disk (all (x, y) with $x^2 + y^2 < 1$). Then int $(A) = A$.

3. (8 points) Consider the curve in the plane given by $\gamma(t) = (1, t)$ if $t < 1$, and $\gamma(t) = (1, 1)$ if $t \geq 1$. Is this curve continuous at $t = 1$? [If yes, tell how to choose δ given ϵ; if no, tell how to choose ϵ.]

4. (20 points) Let X, d be a metric space, and let p be a point of X. Let A be the subset of X consisting of all points q of X with $d(p, q) < 1/2$. Which of the following are *always* true (i.e., no matter what X, d, and p are) [Please answer in a sentence or two.]:

   i) int $(A) = A$

   ii) bnd (A) is the set of all points q with $d(p, q) = 1/2$

   iii) A is bounded

   iv) A is connected

5. (10 points) Let X, d be a metric space, and let A be a bounded subset of X. Then bnd (A) is bounded. [Give a formal proof.]

Test 2 – Solutions

1. Let X be the set consisting of just one point, p and let $d(p, p) = 0$. This is a metric space. The subset of X consisting of just p, and the empty subset of X, are clearly both connected. Thus, every subset of X is connected, and so certainly every bounded subset of X is connected.

2. i) True. Every point of int (X) is certainly in X. For the converse, let p be a point of X. Then every point of X within distance 231 of p is in X, and so p is in int (X). ii) True. X is the union of A and A^C, while the union of two bounded sets is bounded. iii) False. Let X consist of two points, p and q, and let $d(p, p) = d(q, q) = 0$, and $d(p, q) = 1$. This is a metric space. Let A consist of just p. Then A^C consists of just q, and so A and A^C are connected. But X is not connected. iv) False. Let X be the plane, let d' be the usual geometrical distance, and set $p = (0, 0)$ and $q = (2, 0)$. For r and s points of X neither p nor q, set $d(r, s) = d'(r, s)$, $d(p, r) = d(r, p) = d'(r, q)$, $d(q, r) = d(r, q) = d'(r, p)$ and $d(p, q) = d(q, p) = d'(p, q)$. This is a metric space! In it, int (A) is all points (x, y) with $0 < x^2 + y^2 < 1$, which is different from A.

3. This curve is continuous at $t = 1$. Given positive ϵ, any δ less than or equal to ϵ will do. Indeed, $d(\gamma(t), \gamma(1))$ is zero if $t \geq 1$, and is $d((1, t), (1, 1)) = 1 - t$ if $t < 1$. So, if $|1 - t| \leq \delta$, and $\delta \leq \epsilon$, then $d(\gamma(t), \gamma(1)) \leq \epsilon$.

4. i) Always true. By a theorem, int $(A) \subset A$. For the converse, let q be in A, say $d(p, q) = a < 1/2$. Choose positive ϵ such that $a + \epsilon < 1/2$. Then any point within ϵ of q is, by the triangle inequality, within $1/2$ of p, and so is in A. Since every point within ϵ of q is in A, q is in int (A). ii) Not always true. Let X be the plane, and let $d(p, q) = 0$ if $p = q$, and $d(p, q) = 1/2$ otherwise. Then A consists only of the point p, and bnd (A) is the empty set. But the set of all points q with $d(p, q) = 1/2$ is all of X except for the point p. So, bnd (A) is not the set of all q with $d(p, q) = 1/2$. iii) Not always true. Let X be the plane, and $d(p, q) = 0$ if $p = q$ and $d(p, q) = 1/4$ if $p \neq q$. Then $A = X$. But X itself is not bounded. iv) Not always true. The example of iii) in which A is certainly not connected will do.

5. Let positive ϵ be given. Then since A is bounded, there is a finite set of points of X, $p_1 \ldots p_n$, such that every point of A is within distance $\epsilon/2$ of at least one of them. Let p be any point of bnd (A). Then some point, say q, of A is within $\epsilon/2$ of p. But now, by triangle inequality, (since $d(p, q) \leq \epsilon/2$ and $d(q, p_i) \leq \epsilon/2$) we have $d(q, p_i) \leq \epsilon$. Thus, we have shown that every point of bnd (A) is within ϵ of at least one these p_i. So, since ϵ is arbitrary, bnd (A) is bounded.

Phy Sc 112   **Test 3**   March 10, 1978

1. (20 points) Let X be the plane, and let $\mathcal{J}$ consist of the empty set, X itself, and the subset of X consisting of all (x, y) with $x > 0$. So, $X, \mathcal{J}$ is a topological space. Let $A \subset X$ consist of all (x, y) with $x^2 + y^2 < 1$.
 i) Find int (A), and explain your answer.
 ii) Find bnd (A), and explain your answer.
 iii) Is A connected? Explain your answer.
 iv) Let X', $\mathcal{J}'$ be the plane with the usual topology, and $X \xrightarrow{\gamma} X'$ the identity mapping. Is γ continuous? Explain.

2. (16 points) true or false? Please explain in a sentence or two.
 i) The intersection of any two open neighborhood of point p in a topological space is an open neighborhood of p.
 ii) The union of any two open neighborhoods of point p in a topological space is an open neighborhood of p.
 iii) The image of an open set under a continuous mapping of topological spaces is an open neighborhood of p.
 iv) The image of a disconnected set under a continuous mapping of topological spaces is a disconnected set.

3. (9 points) <u>Theorem</u>. Let $X, \mathcal{J}$ be a topological space, and A and B open subsets of X having no point in common. Let $A \subset$ bnd (B). Then A is empty. [Give a formal proof.]

4. (15 points) A topological space $X, \mathcal{J}$ is said to be *locally connected* if every point p of X has a connected open neighborhood. Is every discrete topological space locally connected? Every indiscrete topological space? The plane with the usual topology? [Please explain why.]

<u>Note</u>: "The plane with the usual topology" is the underlying topological space of the plane with usual geometrical distance.

Problem Set 1 Due: January 16, 1978

1. Draw a picture of the set of points (x, y) in the plane satisfying $xy = 1$.

2. Find a simples characterization of each of the following two sets in the plane: i) those (x, y) satisfying $x^2 = y^2 \leq 0$; ii) those (x, y) such that $x \geq 0$ or $y \geq 0$ or $x + y \leq 0$

3. Which of the following describe sets in the plane? i) all (x, y) such that, for some y, $x = y$; ii) Start with the three points $(0, 0)$, $(0, 1)$ and $(1, 0)$. Then include any point lying on a line joining any two of these. Then include any point lying on a line joining any two of the resulting points. etc. iii) all (x, y) such that $x + y + z = 3$.

4. We have given examples of various statements which do, and various which do not, define sets in the plane. Find the best example you can of a statement which is borderline between the two.

5. Consider the following two properties of sets in the plane: i) "is a one-dimensional set", and ii) "the set includes the point $(0, 0)$". Which, in your opinion, will ultimately be topological properties.

6. Consider the "disk with a hole in the center", i.e., the set in the plane given by $x^2 + y^2 > 1$ and $x^2 + y^2 < 1$ What do you think its boundary and interior are? Do you think it is connected? bounded?

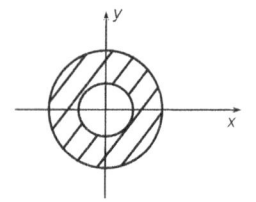

Problem Set 1 – Solutions

1. This set consists of two hyperbolas, as shown in the figure.

2. i) Neither x^2 nor y^2 can be negative. So, we could have $x^2 + y^2 \leq 0$ only if $x^2 + y^2 = 0$, i.e., only if both x^2 and y^2 are zero, i.e., only if both x and y are zero. So, this is the set consisting only of the point $(0, 0)$. ii) In order that (x, y) not be in the set, all three of $x \geq 0, y \geq 0$, and $x + y \leq 0$ must fail. That is, x must be negative, y must be negative, and $x + y$ must be positive. But this is impossible. So, this set is the entire plane.

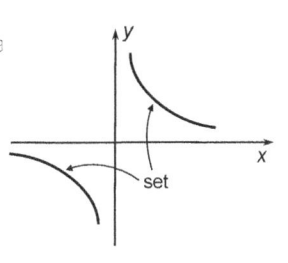

3. i) This does not describe a set. We are to be given (x, y), so what does "for some y" mean? ii) This does describe a set, and in fact it is the entire plane. iii) This does not describe a set. What is this z?

4. Your guess is as good as mine.

5. i) This problem will be a topological property, since it would seem to be unchanged under stretching and pulling. [In fact, it is, but it is a very complicated one to state.] ii) This probably will not be a topological property, since stretching and pulling could move a set away from the origin. [In fact, it is not.]

6. I would have guessed that its boundary will be the two circles $x^2 + y^2 = 1$ and $x^2 + y^2 = 4$; its interior will be the set itself; and that it will be both connected and bounded. [In fact, these are precisely the conclusions one would draw by applying our definitions.]

Problem Set 2   Due: January 26, 1978

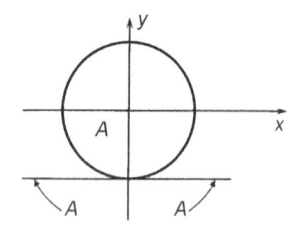

1. Let A be a "disk sitting on a line", i.e., the set of all points (x, y) with either $y = -1$ or $x^2 + y^2 < 1$. Find bnd (A) and int (A). Is A bounded? connected?

2. Find an example of a set which is both bounded and connected; one neither bounded nor connected; one bounded but not connected; one connected but not bounded.

3. Is the interior of every connected set connected? The boundary of every bounded set bounded?

4. Listed below are some proposed modifications of our definitions. In each case, apply the modified definition to the disk (i.e., for "boundary" or "interior" find, with the new definition, the boundary or interior of the disk; for "bounded" or "connected" determine, with the new definition whether the disk is bounded or connected). i) In the definition of "boundary", omit the word "positive". ii) In the definition of "boundary", change "... and also a point ..." to "... or a point ...". iii) In the definition of "interior", change "... for some positive number ..." to "... for any positive number ...". iv) In the definition of of "interior", omit the word 'positive". v) In the definition of "bounded", change "... some point p of the plane and some positive number c" to "every point p of the plane there exists a positive number c such that ...". vi) In the definition of "disconnected", omit "... and no point of A is in bnd (B)."

5. Find all sets A such that int (A) is the entire plane.

Problem Set 2 – Solutions

1. Bnd (A) consists of all points (x, y) with $x^2 + y^2 = 1$ or $y = -1$. Int (A) consists of all points (x, y) with $x^2 + y^2 < 1$. This set is not bounded, but it is connected.

2. A disk is both bounded and connected; two parallel lines are neither bounded nor connected; a set consisting of two points is bounded but not connected; the entire plane is connected but not bounded.

3. It is not true that the interior of every connected set is connected. Let, for example, A consist of two disks together with a line as shown. Then A is connected, but its interior is not. the boundary of every bounded set is bounded.

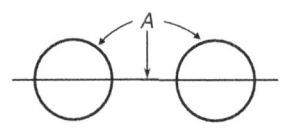

4. i) The boundary of the disk would be the empty set (since, for *every* ϵ one would have to find point q such that $d(p, q) \leq \epsilon$. But what about $\epsilon = -17$?) ii) The boundary of the disk would be the entire plane (since there will always, for any p, be either a q satisfying the first or a q satisfying the second). iii) The interior of the disk would be the empty set (for this is to hold for *every* positive ϵ. What about $\epsilon = 137$?) iv) The interior of the disk would be the entire plane (for the condition is satisfied for *some* ϵ, namely $\epsilon = -13$.) v) The disk would still be bounded. vi) The disk would be disconnected (choosing for C, say, all (x, y) with $x > 0$).

5. Every point of int (A) is a point of A. So, if int (A) is the entire plane, A must be the entire plane.

Problem Set 3   Due: January 30, 1978

1. Find all sets A in the plane such that bnd (A) is the empty set.

2. find all sets A in the plane every subset of which is connected.

3. Find a set A such that int (A) is the disk (all (x, y) with $x^2 + y^2 < 1$) and bnd (A) is the complement of this disk.

4. Is every set in the plane a subset of some connected set?

5. If bnd (A) is connected, need A be connected?

6. Can two different sets in the plane have the same boundary? The same interior?

7. <u>Theorem</u>. Let A be a bounded set in the plane. Then A^C is not bounded. [Give a formal proof.]

8. <u>Theorem</u> Let A and B be sets in the plane, with int $(A) = A$ and int $(B) = B$ Then int $(A \cap B) = A \cap B$. [Give a formal proof.]

9. Find disconnected sets A and B in the plane such that both $A \cup B$ and $A \cap B$ are connected.

10. Think of a subset A of the plane as being "dense" if every point of the plane is "intermingled" with the points of A; if no point of the plane is "out of direct contact" with the set A. [For example, the set of all (x, y) with x rational is to be "dense".] Find a suitable definition of "dense", and state and prove a few theorems.

Problem Set 3 – Solutions

1. The entire plane and the empty set have empty boundaries. Suppose that A were some other set with empty boundary. Then some point of the plane would be in A (since A is not empty), some point of the plane would be not in A (since A is not the entire plane), and no point of the plane would be in bnd (A) (since bnd (A) is empty). That is, we would have that the plane is disconnected. But the plane is connected. So, the only such A are the entire plane and the empty set.

2. If A contained two or more points, than one could form a subset of A consisting of just two points, and this subset would be disconnected. So, A must be either the empty set or a set with just one point.

3. Let A be the set of all points (x, y) in the plane such that either $x^2 + y^2 < 1$ or x is rational.

4. Yes. Every set in the plane is a subset of the entire plane, which is connected.

5. It need not. Our "two squares" set on page 22 has connected boundary, but is not connected.

6. Different sets, same boundary? Sure: For any set A in the plane, A and A^C are different, but have the same boundary, Different sets, same interior? Yes: Let A be a disk, and B a disk together with one additional point.

7. For contradiction, let A be bounded (so there is a point p and a positive number c such that every point of A is within c of p) and let A^C be bounded (so there is a point p' and a positive number c' such that every point of A^C is within c' of p'). Set $a = d(p, p')$. Then every point of A is within c of p, and so within $c + c' + a$ of p. every point of A^C is within c' of p', and so within $c' + a$ of p. and so within $c + c' + a$ of p. So, every point of the plane is within $c + c' + a$ of p: a contradiction.

8. [This is just a special case of the second theorem on page 30. It was a misprint: I meant the union,] i.e., int $(A \cup B) = A \cup B$. For this, a proof would be as follows.] That $A \cup B \subset$ int $(A \cup B)$ follows from the theorem on page 32. That int $(A \cup B) \subset A \cup B$ follows from the theorem on page 26.

9. Let A consist of all (x, y) with $0 \leq x \leq 1$ or $2 \leq x \leq 3$, and let B consist of all (x, y) with $1 < x < 2$ or $3 < x < 4$. Then each is disconnected, while $A \cap B$ is the empty set (so, connected), and $A \cup B$ is a "wide vertical strip" (so, connected).

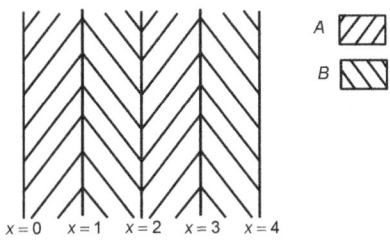

10. A set A in the plane is said to be *dense* if, for every point p of the plane and every positive number ϵ, there is a point q of A with $d(p, q) \leq \epsilon$. Typical theorems include: i) Let A and B be sets in the plane, with $A \subset B$ and A dense. Then B is dense. ii) The

union of two dense sets in the plane is dense. iii) Let A be a dense set in the plane. Then int $(A) \cup$ bnd (A) is the entire plane.

Problem Set 4 Due: February 6, 1978

1. The curve given by $\gamma(t) = (t, t^2)$ is continuous at $t = 0$. For each of the following ϵ-values, find a positive number δ such that, whenever $|t - 0| \leq \delta$, $d(\gamma(t), \gamma(0)) \leq \epsilon$, $\epsilon = 3; 1/2; 1/100$.

2. The curve given by $\gamma(t) = (0, 0)$ if t is rational, $\gamma(t) = (1, 0)$ if t is irrational is not continuous at $t = 1$. Set $\epsilon = 1/2$. Show that there is no positive number δ such that, whenever $t - 1| \leq \delta$, $d(\gamma(t), \gamma(1)) \leq \epsilon$.

3. Let γ be a continuous curve in the plane. Define a new curve, $\hat{\gamma}$, as follows. For t a number, let (x, y) be the point $\gamma(t)$, and then set $\hat{\gamma} = (x+1, y)$. That is, $\hat{\gamma}$ is just γ "displaced one unit to the right along the x-axis". Show that $\hat{\gamma}$ is continuous. [Hint: Fix t_0. Since γ is continuous, for every positive ϵ there exists a positive δ such that ... for γ. Now let ϵ be a positive number. We must find positive $\hat{\delta}$ such that ... for $\hat{\gamma}$. What should we choose for this $\hat{\delta}$?]

4. Let X be the plane. For p and q points of X, set $d(p, q)$ = geometrical distance from p to $q + 1$. Is X, d a metric space? Why?

5. Let X be the plane. For p and q points of X, let $d(p, q)$ be the area of the triangle with vertices p, q and the origin, Is X, d a metric space? Why?

6. Let X be the set of positive integers. For p and q points of X (i.e., positive integers), set $d(p, q) = |p/q - 1|$. Is X, d a metric space? Why?

7. Let X, d and X, d' be metric spaces (i.e., two distances for the same set X). For p and q any two points of X, set $d''(p, q) = d(p, q) + d'(p, q)$. Is X, d'' a metric space? Why?

8. Let X be the plane. For p and q points of X, let $d(p, q)$ be the length of a side of the smallest square, with vertical and horizontal sides, which contains both p and q. Is X, d a metric space? Why?

144   APPENDIX .

Problem Set 4 – Solutions

1. The following are approximately the largest δ's which do the job (i.e., any smaller δ's are ok): 1.6, .45, .0009.

2. Since "1" is rational, $\gamma(1) = (0,0)$. Let some positive ϵ be given. Then there is some irrational number t such that $|t - 1| \leq \delta$. But, for this t, $\gamma(t) = (1,0)$, and so $d(\gamma(t), \gamma(0)) = 1$, which is not less than or equal to $1/2$. Thus, for every positive δ, we have found a t with $|t - 1| \leq \delta$, but such that $d(\gamma(t), \gamma(0)) \leq \epsilon$ fails.

3. Fix t_0, and let positive ϵ be given. Then, since γ is continuous at $t = t_0$, there is a positive δ such that, whenever $|t - t_0| \leq \delta$, $d(\gamma(t), \gamma(t_0)) \leq \epsilon$. But, by definition of $\hat{\gamma}$, we have $d(\hat{\gamma}(t), \hat{\gamma}(t_0)) = d(\gamma(t), \gamma(t_0))$. So, whenever $|t - t_0| \leq \delta$, $d(\hat{\gamma}(t), \hat{\gamma}(t_0)) \leq \epsilon$. We have thus shown that $\hat{\gamma}$ is continuous at $(t = t_0)$. Since t_0 is arbitrary, we have shown that $\hat{\gamma}$ is continuous.

4. No, $d(p, p) = 1$, not zero, so the first condition fails.

5. No. Let p, q, and the origin be arranged as shown. Then the triangle reduces to a line, and $d(p, q) = 0$, while $p \neq q$. The first condition fails.

6. No. It is not true that $d(p,q) = d(q,p)$ (e.g., for $p = 1$ and $q = 2$, so $d(p,q) = 1/2$ while $d(q,p) = 1$). Second condition fails.

7. Yes. each of the three conditions for d'' follows immediately from the corresponding conditions for d and d'.

8. Yes. the first two conditions are immediate. To check the third condition, one just has to play around with squares for a while to see that it is true.

Problem Set 5 Due: February 13, 1978

1. let X, d be a metric space. For p and q any two points of X, set $d'(p, q) = d(p, q)$ if $d(p, q) \leq 1$, and $d'(p, q) = 1$ if $d(p, q) > 1$. Show that X, d' is a metric space.

2. Let X, d be a metric space. Let X' be a set consisting of the points of X together with one additional point, x. For p and q in X, set $d'(p, q) = d(p, q)$; for p in X, set $d'(p, x) = 1$; set $d'(x, x) = 0$. Is X', d' a metric space? Why or why not?

3. Let γ be a continuous curve in the plane. Consider a new curve in the plane, $\hat{\gamma}$, given by: for t any real number, $\hat{\gamma}(t) = \gamma(2t)$. Thus, $\hat{\gamma}$ represents "the same path as γ, but traversed twice as fast". Show that $\hat{\gamma}$ is continuous.

4. Consider the following candidate for a set having more elements (as we have defined it) than the set of integers, but fewer elements than the set of numbers between 0 and 1: the set of numbers between 0 and 1/2. Why does this set not work?

5. Let S, S', and S'' be sets, and suppose that S has the same number of elements as S', and S' the same number of elements as S'' (as we have defined "same number of elements"). Show that S has the same number of elements as S''.

Problem Set 5 – Solutions

1. Since the first two conditions are immediate, we have only to worry about the triangle inequality, $d'(p,q)+d'(q,r) \geq d'(p,r)$. Since in this metric space no distances exceed one, this could fail only if the left side were less than one (since certainly the right side cannot greater than one). But if the left side is less than one, we have $d'(p,q) = d(p,q)$ and $d'(q,r) = d(q,r)$ (since each is less than one), and so, by the triangle inequality in X, d, that $d(p,r)$ is also less than one. Hence, we must also have $d'(p,r) = d(p,r)$. Our triangle inequality now follows from that inequality in X, d.

2. This is not in general a metric space. Let p and r be points of X with $d(p,r) = 5$, say. Then the triangle inequality would demand $d(p,x) + d(x,r) \geq d(p,r)$, which fails.

3. Fix number t_0, and let positive ϵ be given. Then, since γ is continuous at $2t_0$, there is a positive δ such that, whenever $|t-2t_0| \leq \delta$, $d(\gamma(t), \gamma(2t_0)) \leq \epsilon$. That is, setting $t' = 1/2$, whenever $|t' - t_0| \leq \delta/2$, $d(\gamma(2t'), \gamma(2t_0)) \leq \epsilon$. But, since $\hat{\gamma}(t) = \gamma(2t)$, this just means that $\hat{\gamma}$ is continuous at t_0).

4. But the set of numbers between 0 and 1/2 has the same number of elements as the set of numbers between 0 and 1. Indeed, a correspondence is the following: With any number x between 0 and 1/2, associate the number $2x$ (between 0 and 1). Every number of one set is made to correspond with some number in the other, and vice versa, and no numbers of either set are left over.

5. We have a correspondence between the elements of S and S' (such that every element of S corresponds to one of S', and vice versa, and such that no elements of S or S' are left over); and also a correspondence between the elements of S' and S'' (etc.). Now consider the following correspondence between the elements of S and S'': Given an element of X, find the element of S' to which it corresponds (under the first correspondence), and then the element of S'' to which that element of S' corresponds (under the second correspondence). One checks that this correspondence between S and S'' has all the required properties.

Problem Set 6 Due: February 20, 1978

1. Let X be the plane, d the usual geometrical distance, and d' twice that geometrical distance. Show that, for any set $A \subset X$, bnd (A) and int (A) are the same using d as d'.

2. Is it true that for A any subset of any metric space, with bnd $(A) = A$, then int (A) is the empty set? Why or why not?

3. Let X be the "circle" metric space of page 47. Find all bounded subsets of X.

4. Let X, d be a metric space, with A having only a finite number of points. Prove that, for any $A \subset X$, int $(A) = A$ and bnd (A) is the empty set.

5. Is it true that, for any metric space X, d and for any subset A of X consisting of just two points, A is disconnected? Why or why not?

6. Let X, d and X, d' be metric spaces such that, for any $A \subset X$, int (A) (with respect to d) = int (A) (with respect to d'). Show that bnd (A) (respect to d) = bnd (A) (with respect to d').

Problem set 6 – Solutions

1. Let int and int' denote interiors using d and d', respectively. For p in int (A) there is a positive ϵ such that every point within d–distance of p is in A. But then every point within d'–distance 2ϵ would be in A, and so p would be in int (A). Similarly (using $1/2\epsilon$ instead of 2ϵ), every point of int' (A) is in int (A). That one also gets the same boundaries is immediate from the fifth theorem on page 68.

2. int (A) must be a subset of A, and can have no points in common with bnd (A). So, if bnd $(A) = A$, int (A) must be the empty set.

3. Every subset of X is bounded, because X itself is bounded. Indeed, given positive ϵ, let $p_1, \ldots, p_n$ be a finite number of points spaced less than ϵ apart around the circle. then every point of X is of course within ϵ of at least one of these points.

4. Let p be in A, and consider the numbers $d(p, q)$ for $q \neq p$. This is a finite number of positive numbers: Let ϵ be smaller than the smallest. Then every point within ϵ of p is in A (for we choose ϵ such that the only such point is p itself). So, p is in int (A). So, int $(A) = A$. Similarly, int $(A^C) = A^C$. Since the points of bnd (A) are precisely those points of X in neither int (A) nor int (A^C), bnd (A) must be the empty set (for every point is in either int $(A) = A$ or int $(A^C) = A^C$).

5. This is true. Let A consist of p and q, with $p \neq q$. Fix a positive number b less than $d(p, q)$, and let B be the subset of X consisting of all points r with $d(p, r) \leq b$. Then, of course, p is in B and q is not. We claim further that no point of A is in bnd (B), i.e., that neither p nor q is in bnd (B). Certainly p is not in bnd (B), for it is not true that there is a point within b of p and not in B, We further claim that q is not in bnd (B). Indeed, choose a positive number ϵ much that $b + \epsilon \leq d(p, q)$. (possible, since b is less than $d(p, q)$). Then no point within ϵ of q could be in B (i.e., within b of p), for were there such a point, r, the triangle inequality, $d(q, r) + d(r, p) \geq d(p, q)$, would contradict our choice of ϵ above. So, q is not in bnd (B). So, A is disconnected.

6. Since, by the fifth theorem on page 68, bnd $(A) = [\text{int}(A)]^C \cap [\text{int}(A^C)]^C$, it is immediate that two metrics having the same interiors also have the same boundaries. [Also, is fact, same connected set.]

Problem Set 7 Due: February 27, 1978

1. i) Let X, d be a metric space, and p and q points of X with $p \neq q$. Show that there is a self-interior set A with p in A and q not in A.

ii) Let X, d be a metric space, and $X, \mathcal{J}$, its underlying topological space, and p and q points of X with $p \neq q$. Show that there is an open set A in this topological space with p in A and q not in A.

iii) Let X be the plane with the indiscrete topology, and p and q points of X with $p \neq q$. Show that there is an open set A in this topological space with p in A and q not in A.

iv) Show that the plane with the indiscrete topology is not the underlying topological space of any metric space.

2. Let X be a set with just three points. Find all $\mathcal{J}$'s (i.e., all collections of subset of X) such that $X, \mathcal{J}$ is a topological space. [Hint: There are exactly twenty-nine of them.]

3. Let X, d be a metric space, with X finite. Show that the the underlying topological space is discrete.

4. Let X be a set, and $\mathcal{J}$ a collection of subsets of X such that the second and third conditions for a topological space are satisfied. Is it true that, if there is now included in the collection $\mathcal{J}$ also X itself and the empty set, then the result will always be a topological space? Why or why not?

Problem set 7 – Solutions

1. i) Let $d(p,q) = a$, a positive number. Let A be the set of all points r with $d(p,r) < a/2$. Then clearly p is in A and q is not in A. By the theorem on page 82, A is self-interior. ii) Since the open sets in the underlying topological space are the self-interior sets, this follows from i). iii) In the indiscrete topology, the only open sets are X and the empty set. But A empty will not do, for p will not be in A; and $A = X$ will not do, for q will be in A. iv) Let X, d be any metric space with X the plane, and let $p \neq q$. By ii), these is an open set, in the underlying topology, containing p but not q: by iii) there is no open set, in the indiscrete topology, containing p but not q. So these topologies must be different.

2. One just has to hunt them down. There is the indiscrete (1) and the discrete (1). The rest are: $[x, y, xy]$ (3) [Which means: the open sets are the empty, X itself, the set consisting of just x, the set consisting of just y, and the set consisting of just x and y; and the total number of such possibilities, when rearrangements of x, y, and z are included, is 3], $[xy](3)$, $[x](3)$, $[x, yz](3)$, $[x, xy](6)$, $[x, xy, xx](3)$, $[x, y, xy, xz](6)$.

3. We have seen that, in a metric space with X finite, int $(A) = A$ for every $A \subset X$, i.e., that every A is self-interior. So, in the underlying topological space, every subset of X is open. That is, the underlying topological space is discrete.

4. This is true. Let $\mathcal{J}'$ denote $\mathcal{J}$, together with $\emptyset$ (empty set), and X itself. 1. X and $\emptyset$ are certainly in $\mathcal{J}'$. 2. Let A and B be in $\mathcal{J}'$. If either is empty, then $A \cap B = \emptyset$, and is in $\mathcal{J}'$. If either is X, then $A \cap B$ is the other, and so is in $\mathcal{J}'$. If neither is empty or X, then each of A and B is in $\mathcal{J}$ and so $A \cap B$ is in $\mathcal{J}$, and so $A \cap B$ is in $\mathcal{J}'$. 3. Let C be the union of any collection of sets in the collection $\mathcal{J}'$. We might as well leave out empty sets in this union, since they will not change C. If any of these sets is X, then $C = X$, and so is in $\mathcal{J}'$. If none is X, then they are in $\mathcal{J}$, and so the union is in $\mathcal{J}$, and so that union, C is in $\mathcal{J}'$.

Problem Set 8 Due: March 8, 1978

1. Let X, $\mathcal{T}$ be a topological space in which every subset of X consisting of a single point is open. Prove that this topological space is discrete.

2. Consider the topological space of pages 78 – 79. Find the connected subsets of X.

3. Let X be the plane, and let the open sets be the empty set and all subset of X containing the origin. Let A consist of all (x, y) with $x^2 + y^2 < 1$.

i) prove that this is a topological space.

ii) Find int (A).

iii) Find bnd (A).

iv) Is A a connected? Why or why not?

4. Let X, $\mathcal{T}$ be a topological space, and A a connected subset of X. Prove that $A \cup$ bnd (A) is connected. [Note that one would have expected this intuitively!] [Hints: Suppose disconnected, so there is a set C such that some point, p, of $A \cup$ bnd (A) is in C, some point, q, of $A \cup$ bnd (A) is not in C, and no point of $A \cup$ bnd (A) is in bnd (C). Now it could hardly be that both p and q are in A (as opposed to bnd (A)), for that would mean that A would be disconnected. Suppose, then, that p is in bnd (A). Now use i) the fact that p is in C, ii) the fact that p is in bnd (C) (since no point of $A \cup$ bnd (A) is in bnd (C)), and iii) the fact that p is in bnd (A) to show that there must be a point p' in A and C. Similarly, if q were in bnd (A), show that there must be a point q' in A and not in C. Thus, we wind up finally with a point (p or p') in A and in C, a point (q, or q') in A and not in C, and with no point in A in bnd (C). But this would imply that A is disconnected.]

Problem set 8 – Solutions

1. In a topological space, any union of open sets is open. But any subset of X can be written as a union of sets consisting of a single point. So, if these are open, the space is discrete.

2. For any subset B of X, either int (B) is empty, or int (B^C) is empty. So how, given A, could one find B such that some point of A is in B, some point of A is not in B, and no point of A is in bnd (B) (since if no point of A is in bnd (B), the point of A in B must be in int (B), and the point of A not in B must be in int (B^C))? Every subset of X is connected.

3. i) 1. The empty set and X itself are open. 2. The intersection of two sets containing the origin contains the origin. 3. The union of any collection of sets containing the origin contains the origin. ii) int $(A) = A$. Indeed, this A is open, since it contains the origin. iii) bnd $(A) = A^C$. Indeed, let p be in A^C. Then every open neighborhood of p contains a point not in A (namely, p), and also a point in A (namely, the origin). iv) In order that A be disconnected, we must find B such that some point of A is in B, some point of A is not in B, and no point of A is in bnd (B). But, if B contains the origin, then bnd $(B) = B^C$ – so, the point of A not in B is in bnd (B); if B does not contain the origin, then bnd $(B) = B$ – so, the point of A in B is in bnd (B). In either case, there would be a point of A in bnd (B). So, we shall never find set B which shows disconnected. This A is connected.

4. See notes, pages, 95 and 96.

About the author

Robert Geroch is a theoretical physicist and professor at the University of Chicago. He obtained his Ph.D. degree from Princeton University in 1967 under the supervision of John Archibald Wheeler. His main research interests lie in mathematical physics and general relativity.

Geroch's approach to teaching theoretical physics masterfully intertwines the explanations of physical phenomena and the mathematical structures used for their description in such a way that both reinforce each other to facilitate the understanding of even the most abstract and subtle issues. He has been also investing great effort in teaching physics and mathematical physics to non-science students.

Robert Geroch with his dog Rusty

Made in the USA
Las Vegas, NV
18 September 2023